G000123546

The Theory of Technological Change and Economic Growth

The last two centuries have witnessed sustained technological innovation which by historical standards has been unique in two respects: it has been unusually intense and its variation between nations and types of economic activity has been remarkably large. This book surveys the empirical evidence and the numerous theories of technological change and economic growth in an attempt to provide a unified empirical and theoretical framework for interpreting this intensity and variation. Part I deals with the invention–innovation–diffusion process at the level of firms and industries. It seeks to develop the microfoundations for the macro-oriented analysis of Part II. Central to that macro analysis are two hat-shaped or bell-shaped relationships. One of these describes the change over time of the innovation rate in the technology frontier area (TFA), the part of the world that is technologically most advanced. The other describes the innovation growth paths in the countries outside the TFA.

The wide empirical microeconomic basis and historical perspective of the book places it in the tradition associated with such economists as Schumpeter and Kuznets. However, the book attaches great value to theoretical insight, and this places it in the modern tradition.

Stanislaw Gomulka is Reader in Economics at the London School of Economics. Educated in Poland, he has taught at the University of Pennsylvania and Aarhus University and has held visiting and research appointments at the Netherlands Institute for Advanced Studies and Columbia, Stanford and Harvard Universities.

The Theory of Technological Change and Economic Growth

Stanislaw Gomulka

I once belonged to Chris Freeman. Where am I now? Tell us at: www.freemanchris.org

ROUTLEDGE

London and New York

To Joanna and Michael

First published 1990
by Routledge
11 New Fetter Lane, London EC4P 4EE

Simultaneously published in the USA and Canada
by Routledge
a division of Routledge, Chapman and Hall, Inc.
29 West 35th Street, New York, NY 10001

© 1990 Stanislaw Gomulka

Printed and bound in Great Britain by Mackays of Chatham PLC, Kent

British Library Cataloguing in Publication Data

Gomulka, Stanislaw
    The theory of technological change and economic growth.
    1. Technological development. Economic aspects
    I. Title
    338.06
    ISBN 0-415-03692-5
    ISBN 0-415-05238-6 pbk

Library of Congress Cataloging in Publication Data

Gomulka, Stanislaw.
    The theory of technological change and economic growth /
    Stanislaw Gomulka
        p.   cm.
    Includes bibliographical references.
    ISBN 0-415-03692-5
    ISBN 0-415-05238-6 (pbk)
    1. Technological innovations—Economic aspects. 2. Economic
development.   I. Title.
HC79.T4G655   1990
338'.064—dc20
                    89-77047
                    CIP

Contents

Part Two   Macroeconomics of Innovation, Technology Transfer, and Growth

Acknowledgements

The original invitation to write this book came from Mervyn King of the LSE, and I owe him much for the advice he offered at the initial, formative stage of the project, and for his subsequent encouragement. The task was difficult to accomplish well: to write a monograph which would deal with central issues of the field and which could also serve as a textbook. A large part of the first draft of the book was written at the Netherlands Institute for Advanced Studies in Wassenaar, and I wish to thank the directors and staff for their help. I have benefited much from the comments of several of my LSE colleagues on the various chapters of the book. I wish to acknowledge in particular the suggestions of David de Meza and Mark Schaffer. Alan Jarvis and Ruth Jeavons of Routledge and Christine Sharrock of Omega Scientific were most helpful with their willingness to accept last-minute changes and for seeing the work through its various publication stages.

ACKNOWLEDGEMENTS

Part one

Microeconomics of invention, innovation, and diffusion

Chapter one

Preliminary concepts and relations

The first three chapters of this book are introductory. They are intended primarily to familiarize the reader with some of the economic concepts that will be needed later in the book. The first such preliminary concept is that of the production sector, which will be presumed to consist of production units called firms or enterprises and to be distinct from the sector consisting of consumption units called households. We shall have little to say about the household sector. The innovation activity and the growth of the productivities of inputs in firms, industries, and national economies worldwide will be our primary interest. The production sector is presumed to supply goods of two categories: 'conventional' goods, such as intermediate and investment inputs and consumer goods, and 'progress' goods in the form of new inventions and new skills intended to enhance the welfare-creating capacities of firms and households. Accordingly, in the production sector we distinguish between the conventional activity and the inventive activity. In our introductory chapters we shall identify and discuss the major characteristics of the nature and size of the inventive activity, and the ways in which the latter interacts with and influences the composition and growth of conventional activity. We shall also identify the major stylized facts about world inventive activity, concerning its size and changes over time as well as the distribution among countries and industries in recent times.

Two major definitional qualifications should be made at the outset. One concerns the distinction between the production sector and the household sector. It should be noted that if the enjoyment which we derive from the consumption of goods or from work is seen as one of the (ultimate) goods that economic activities provide, then households themselves should be viewed as production units, especially since in addition to enjoyment they also supply labour services, spiritual experiences, and recreation, as well as conventional goods of the do-it-yourself variety. This is clearly a valid point; it actually led William Nordhaus and James Tobin in 1972 to propose a wider measure of the net effect of economic activity than the widely used net national product (NNP). Their so-called measured economic welfare (MEW) concept attempts to take proper account, in addition to NNP, of the value of leisure, the value of the non-market services of the household sector, and the environmental costs of production, among other things. In our analysis of technological innovation and economic growth it is not conceptually essential to exclude the household sector completely from the production activity. It is only for the well-known reasons of statistical convenience or necessity that the measures of national product that are commonly used exclude the contribution of the household sector. We shall also use these measures and, to that extent, we shall usually restrict the notion of the production activity to its conventional meaning.

The 'black' and 'parallel' economy is another part of the production activity where there are serious measurement problems. We shall make the convenient assumption that it is a constant fraction of the total production activity. However, if the shifts in the distribution of resources between this and the observable sector were large they could seriously disturb the quality of the published output and productivity data, something which some economists suggest actually took place in the 1970s. We have to keep this possibility in mind when using the data, and only if reliable information permits take account of the unobservable economy in our attempts to measure the innovation rate and economic growth of the total production activity.

Production processes, techniques, and technology

Production of goods in any individual enterprise involves combining different kinds of primary inputs, such as unskilled labour and natural resources, with intermediate inputs, such as semi-fabricated materials and energy, and with the services of skilled labour and fixed capital. Production is often a very complex operation, but it can be broken down into many distinct standard operations, some or all of which may take place simultaneously. These are called *production processes* or *activities*.

Following Leontief (1947) and Morishima (1976), we can represent the whole system of processes for each enterprise by a tree. In the tree shown in Figure 1.1, there are four processes. In process (i) inputs 1, 2, and 3 are combined to produce good 8, in short $(1, 2, 3) \rightarrow (8)$. There are two alternative methods of obtaining good 9: (ii) $(3, 4, 5) \rightarrow (9)$ and (iii) $(5, 6, 7) \rightarrow (9)$. In

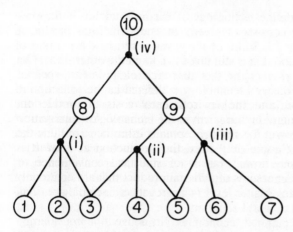

Figure 1.1 Geneology of production. An enterprise with four processes and two alternative methods of producing one final good.

process (iv), which represents the next stage of production, goods 8 and 9 are themselves used as inputs to produce 10, the final output of the production unit. Since product 9 can be obtained by employing either process (ii) or process (iii), two alternative methods of producing good 10 are available to our firm in Figure 1.1, one involving processes (i), (ii), and (iv), and the other involving processes (i), (iii), and (iv). These different methods of producing the final good or goods are called *techniques,* and the set of all techniques available to a firm is its *technology* in the narrow sense.

A process may also require a management input, and this requirement would be reflected in the process's list of inputs. When several processes are involved, we may need someone in the firm to know that they are in fact available and to be able to make the best selection from among them. We include this higher level of organizational and management knowledge in our concept of technology in the broader sense, or simply technology. Consequently, our technology set includes not only purely technological processes reflecting different ways of combining inputs, but also organizational processes reflecting different ways of combining the processes themselves. The outputs of organizational processes are services which may not be sharply defined. For conceptual convenience we can imagine that to each technique of production there corresponds one organizational process with the necessary services of the firm's management and related staff and other resources as its inputs but with zero outputs.

How do we define technology at the industry level? Suppose that an industry consists of several independent firms producing the same final good. Some of the firms may know a method of producing the good that is still unknown to all the other firms. Our definition must respect the fact that technology is firm-specific. Therefore the industry's technology is defined as the collection of all the firm-specific technology sets, each of which contains the production techniques of one firm in the industry. An enlargement of the technology set for any firm, even if it has occurred through the interfirm diffusion of the existing technological knowledge within the industry, would then also represent an enlargement of the industry's technology set. We adopt a similar definition of technology for any higher level of aggregation, including national economies and the world economy.

In centrally planned economies firms have a considerable degree of autonomy, but not full organizational independence. A group of firms is typically organized into an association, associations are organized into a sector headed by a ministry, and sectors are organized into an economy headed by the government. It is a hierarchical structure with each level capable of influencing other levels, especially those below. The association's technology set must therefore be defined to include, in addition to the firm-specific sets, a process or processes pertaining to the organizational activity of the association's head office. This activity could be conducted in a number of ways, and any choice of conduct would be reflected in the quantities of the final outputs and inputs of the association as a whole. The same procedure can be used in the treatment of individual ministries and the government as a whole.

Efficient techniques and technological progress

Whatever the level of aggregation, any enlargement of the corresponding technology set represents, by definition, a *technological change*. However, not every enlargement of the technology set represents what we should like to call technological progress. Inefficient techniques – those which with the same inputs produce less of one or more outputs, or those which require a greater quantity of inputs to produce the same outputs as some other techniques – should not be selected for use whatever the prices of the relevant inputs and outputs, with the exception of cases in which some of the inputs can, like air, be obtained free of charge. Any addition of an inefficient technique to a firm's technology set would, in our terminology, represent zero technological progress for that firm. We define any technique that is not inefficient as efficient, and we say that *technological progress* takes place when a firm's subset of efficient techniques is enlarged or when the newly arrived technique(s) dominates one or more of the existing efficient techniques so that the latter become inefficient. If a firm replaces one efficient technique by another, we cannot be certain that technological progress has taken place, except in the obvious case when the old technique would become inefficent after the acquisition of the new one.

Production processes involve many inputs and one or more outputs. If a process requires that the proportions in which the inputs to be used and the outputs to be produced are fixed, then the process can be fully described by the underlying input–output coefficients. We simply select one of the outputs and take the quantity of the selected output as a measure of the scale on which the process is to be operated. This scale is sometimes referred to as the intensity of the process. Therefore the input–output coefficients for the process in question are the quantities of the inputs and the other outputs per unit of that particular scale or intensity.

However, in practice these input–output coefficients are rarely fixed. To begin with, the rates at which workers operate their machinery and the intensity of their work are rarely fully technologically determined, but are instead a product of conventions or a subject of negotiations between workers and management. As such, these rates are inevitably dependent on particular traditions, institutions, and motivations of all kinds. The resultant input–output coefficients in part mirror the human environment in which the production process takes place. A process which is efficient in one environment may thus be inefficient in another. This is one reason why technological change itself is also culturally dependent. There are many other reasons. In fact there are good grounds for suspecting that this interdependence between technological change and the cultural, as well as the institutional, characteristics of a nation is one of the most powerful causes of the observed wide variation in the rates of innovation and economic growth among nations and, to some extent, over time. We shall refer to this interdependence many times throughout the book.

The other major parameter giving rise to variation in input–output coefficients is the scale on which a production process is operated. Processes need not be scale-specific, but neither can they be operated on any arbitrary scale. Usually the description of a process includes also the specification of the range of scales on which it is most efficient to use it. One form of technological progress has been precisely the invention of large-scale processes to supply goods at a low cost for large-volume markets. Indeed, the shift towards a large-scale technology has been one of the dominant trends of the world's industrial revolution, until now at any rate, partly in response to the enlargement of markets: from local to regional, to national, and now to worldwide. The advent of microelectronic devices and sophisticated robots may end or even reverse this trend. Yet it would still remain true to say that a change in the scale of operation may often turn an inefficient process into an efficient one, and vice versa. Consequently the rate of assimilation of a particular technique, and hence the rate of technological progress, may also depend on the size of the market which an industry supplies.

If the number of all products is finite, and for practical purposes this is what we usually assume, the maximum output that a firm can obtain from different combinations of inputs forms, in the product space of all inputs and outputs, a multidimensional sphere called the firm's *production possibility frontier* or production function. The addition of an efficient technique moves this frontier or function outwards, and thus any such movement signifies the presence of technological progress within the firm.

Allocative efficiency, X-efficiency, and relative rationality

A switch from one technique to another involves a change in the composition of production processes. Thus, from a purely engineering point of view, the switch represents a change in technology. However, if both techniques were known to the firm before the change took place, then, in our terminology, they would belong to the same technology set and, as the switch merely represents a movement within the set, no technological change would have been involved. Technological change would take place if the new technique represented an addition to those already in the technology set. Moreover, the change would represent technological progress if the new technique were efficient.

In our quest for clear and consistent terminology care must be taken to see to it that we do not end up with a straitjacket structure of concepts incapable of dealing with interesting real-life situations. Such a situation may arise when techniques which are actually operated become, as experience accumulates, better known to the operating workers. If this increasing familiarity and accumulating experience are mirrored in changing input–output coefficients, these techniques should be presumed as changing over time. Such experience is usually technique related; it will be partly lost if the firm makes a switch from one technique to another. We therefore retain the distinction between moving within a production possibility set and shifting to a new set, but regard changes in experience with what are essentially the same techniques as a component of technological change.

Technological improvement should be clearly distinguished from a purely *allocative improvement*. The latter takes place when no new knowledge is available to the firm and yet it increases the output of one or more of the goods produced, without decreasing the output of any other goods, by re-allocating inputs between processes away from a more to a less inefficient technique, or indeed to an efficient technique. Allocative improvement at the industry level may take place as a result of a re-allocation of inputs among firms, in addition to any intrafirm re-allocations. Allocations which cannot be further improved are called *Pareto efficient*. From this definition it follows that, for any level of aggregation, the production possibility frontier is the set of Pareto-efficient allocations pertinent to that level. If, given the prices and the aim of the firm, an allocation chosen is Pareto efficient but not optimal, there would be scope for a further allocative improvement. As there could not be any 'rational' reason for choosing a Pareto-inefficient technique, the improvements of the first category are usually ignored, and the concept of allocative efficiency is usually reserved for the choice of optimal technique.

An interest in the 'irrational' choices of Pareto-inefficient techniques has been revived by Leibenstein (1966, 1975, 1979), who calls it X-inefficiency. The important phenomenon of such seemingly irrational economic choices can be rationalized by reference to the presence and influence of the non-economic environment. It has already been noted that the input–output coefficients, which we use to define production processes, typically respond to changes in the cultural and institutional environment in which the processes are operated. It is well documented that the intensity with which workers work varies significantly during each day and between days in a particular firm, between firms in a particular country, and between countries. There could also be powerful intrafirm differences. Since the reasons for this variation are complex and not always well understood, it is fitting to call them the X-factor. A firm in which the workers are, for example, badly motivated and produce less than they otherwise could have is then said to be X-inefficient. When this low intensity of effort adversely affects the production of conventional goods only, the firm is statically X-inefficient, but when it is of consequence for the rate of the innovation process itself, the firm is dynamically X-inefficient. In principle, each type of X-inefficiency is independent

9

of the other; a firm or a country may display a high static X-efficiency and a low dynamic X-efficiency. A major effort may also be misguided and consequently result in low values of both static and dynamic X-efficiency, as in China during the Great Leap Forward of the late 1950s.

The reader may have noticed that if, at a meta-level of this conceptual analysis, we rationalize in a similar manner the actual social choices of motivation incentives, attitudes to work, and institutions, by reference to the most basic characteristics of the society and its individual members, then any choice that has been made would be tautologically rational or optimal, and, given the people as they are and their knowledge at the time, any allocation actually made would be efficient. Rationality and efficiency would then lose their absolute and normative meaning. Yet some choices can be judged to have led to inferior economic performance compared with what might have been the outcome if different feasible choices were made instead. Rationality and efficiency would thus still retain their relative meaning. We shall use this concept of relative efficiency in interfirm and intercountry comparisons. We maintain that it is meaningful to say that, for example, the German economy is more X-efficient than the UK economy, although we shall refrain from saying that the latter is X-inefficient. Simply, the United Kingdom's 'optimal' choice implies poorer economic performance than does the German choice. However, we should like to investigate, whenever we can, the underlying cultural, institutional, and other causes.

Climate and geographical conditions are often powerful factors also. In cold and windy Siberia the factory structures have to be more solid and the input of heat greater than in sunny California. Essentially the same method of mining coal or extracting oil would be characterized by different input–output coefficients in different geological environments, and so forth. It is quite obviously the case that the input–output coefficients of a production process depend not only on its purely technological content, but also on the human and physical environment in which the process is or may be operated. Any change in this environment is alone potentially capable of shifting the production possibility frontier. In practice it is difficult, if not impossible, to separate out a shift of this category from a shift due to a technological change proper. However, since the factors underlying the two kinds of shifts are qualitatively very different, it is useful for analytical, if not practical, purposes to maintain this distinction. The environmental changes may improve as well as worsen the production possibilities of a firm. Although the resultant shifts may look like random disturbances of the firm's production possibility frontier, occasionally they could add up to a significant component of the trend movement of that frontier. However, since there is little to be gained analytically by looking into these matters, in what follows the term technological change is reserved to mean an addition of a new technological knowledge, i.e. the technological change proper as defined in the previous section.

Invention, innovation, and the role of science

Production of conventional goods is sometimes presented as an ordered sequence of 'stages', beginning with the extraction of minerals and other 'primary activities', proceeding through the production of intermediate goods of increasing degree of fabrication, and ending up with the stage of producing final goods. The input–output analysis, in contrast, emphasizes the strong interdependence between most production activities. However, the research and development (R&D) sector is much less structured than the conventional sector. Still, a dual characterization of the type above also has merit with reference to R&D activities. Fritz Machlup (1962: 179–81) and Edward Ames (1961) draw a parallel between the degree of fabrication for conventional goods and the closeness to economic application for ideas. The research part of the R&D sector includes basic and applied sciences of widely varying degrees of abstraction, and is concerned mainly with the discovery of new facts or principles about the natural and social worlds. Invention, in contrast, is a new product or production process, with the latter including a new method of organizing economic activity. Inventions in their first prototype form usually undergo a lengthy and costly process of improvement before they are ready for commercial application. An invention may never be adopted by any producer, but when it is an innovation takes place. It follows that innovation should be understood to mean both an act of qualitative change of an economy when a new product (process) starts to be produced (used), and the product (process) itself. Typical outputs of the inventive work are memoranda, working models, or sketches. The development stage, which in principle is intermediate between invention and innovation, results in blueprints, specifications, samples of new products, and pilot plants.

In practice the interrelation between the various stages is usually much more complex than the sequence hypothesis implies. (This hypothesis was modelled by Gomulka (1970); for more details see Chapter 10.) The R&D structure is more of a multidimensional

11

network type than a one-dimensional sequence type. While modern inventions are often inspired, or indeed permitted, by advances in pure science, most inventions are improvements and ramifications of a relatively few major ideas, and apparently many of these ideas have been discovered by the inventors themselves. Jewkes *et al.* (1958), in their classical history of major inventions, provide a rich body of evidence suggesting the presence of a to-and-from stimulus between scientists and inventors. Perhaps the most startling, and symbolic, is the case of the high-pressure steam engine which was invented in the fourth quarter of the eighteenth century, well before the relevant science existed, but then itself led to the enunciation of the law of thermodynamics. However, Jewkes *et al.* note, about a century later Diesel's attempt to apply these laws led to the diesel engine. In another study in this branch of economic history Jacob Schmookler (1966) identifies no simple instance of a scientific discovery initiating an important invention immediately, and in the few instances when he suspects that such initiation stimulus was identified with certainty, it was either an accident or an economic gain. However, the 'immediately' in this statement is important, for there have been many cases found by Schmookler and his associates in which a scientific discovery made at some time in the past was a precondition of an important invention.

We are not going to review here the evidence on the sources of inventions in general, and on the potency or otherwise of scientific discoveries in particular. In any case, the evidence that we have at the moment is not complete enough to permit definitive conclusions, especially with reference to the nineteenth century and earlier. One conclusion that can be drawn is that, although advances in world science are largely, though not completely, independent of economic considerations, these advances do create a potential for invention, different in different fields, and that the exploitation of this depends strongly, especially in the most advanced countries, on the size and effectiveness of the applied oriented R&D effort. In turn, decisions regarding the allocation of this effort between the different fields is strongly dependent on the evaluation, by the R&D and entrepreneurial staff, of the expected economic gain.

This conclusion appears to be particularly well grounded in more modern times. Up to and including most of the nineteenth century, new inventions tended to derive mainly from the more practical arts and mechanical science. However, describing the history of a number of patents that were granted worldwide during the period 1916–45, Alfred Stafford (1950; cited by Schmookler

1966: 39–40) notes the presence of a major shift in inventive output away from the empirical crafts towards the science-based fields of chemistry and electronics. The shift clearly reflects the quest to exploit the vast potentials for profitable invention which were created by chemists and physicists during the second half of the nineteenth century and the early twentieth century. This trend towards science-based innovation has been sustained, and probably even emphasized, in the course of the twentieth century, with further substantial advances being made in fundamental sciences, particularly perhaps in solid state and nuclear physics. An instructive illustration of this strongly science-based character of the modern inventive process is provided by Christopher Freeman (1982) in his account of the inventive activity in the now extremely important field of synthetic materials. Freeman (1982: 48–70) also notes, with reference to the same field, a sharp rise in the share of patents awarded to professional R&D organizations within firms, especially large ones, in comparison with the private inventor. There is little doubt that in the course of the twentieth century world R&D activity has become an industry which is largely science based and professionally run.

Product and process innovations

Product innovation usually refers to the introduction of significantly new products, while process innovation takes place when significantly new ways of making or using existing goods are adopted. It is useful to regard the term 'new' as being firm-specific, so that interfirm diffusion of existing goods or processes is also a form of innovation. Moreover, old products of a new design or different reliability and performance must also be regarded as new products, and new methods of industrial organization, marketing, and management must be regarded as new processes. However, the difference between a new product and a new process is more clear cut at the inventive than at the innovative stage. The reason is that the production of a new good, whether substantially new or not, would almost always require a new combination of inputs and hence be tied up with a process innovation. In turn, a new process very often changes the quality characteristics of the outputs produced, thus resulting in product innovation. Even when it seems that an innovation is clearly identified as essentially of the process category or of the product kind, it is tempting to downgrade the analytical value of the distinction on the grounds that it rests in part on who uses the invention. A new or improved product of one firm may be a process innovation for the firms using the product. This is indeed often, though not always, the case. However, while the distinction between the supplier and the user may be irrelevant when analysis is conducted at a sufficiently high level of aggregation, it is usually of key importance at the level of individual firms. In devising its R&D and investment strategies, and in forecasting its future productivity and market performance, each firm cannot overlook the fact that, while it has some control over its product range, new methods of producing its products may depend largely on the R&D activities of its suppliers. The meaning of the distinction between process and product innovation derives mainly from this fact. Agricultural farms are an example that is particularly simple and yet telling. Ever since the emergence of a large industry supplying the farms with agricultural inputs, most of the R&D effort designed to invent new cost- and effort-reducing methods of producing agricultural output has been undertaken not by the farms themselves but by the industrial firms supplying the farms' inputs. The aim of agricultural R&D, apart from inventing new varieties of agricultural produce, is largely limited just to testing the usefulness of the inventive output arriving from the outside. Yet it is the outside inventive output that has revolutionized the organizational structure of the farms and drastically changed the productivity of the farmer and his land. The relationship between textile machinery producers and textile processing mills is another illustration of the same points. These are two different industries; the mills are very much on the receiving end of the R&D effort of the textile machinery industry. The industry's new machinery inventions have, over time, enabled the processing mills not only to reduce costs but also to introduce new textile products. There are fields of economic activity where process and product innovations are not as clearly separated as in the two examples above. There are also industries where both kinds of innovation take place largely within the same industry or even within the same firm. Chemicals used to be one such industry, but a change has been taking place in the post-war period and recent evidence suggests that 'the majority of plants are now engineered and built by process plant contractors' (Freeman 1982: 41).

The generalization that the foregoing discussion suggests is that the rate of process innovation in any industry is largely determined by the rates of product innovation in the supplying industries, with the rates possibly weighted by the shares of supplies in the total product inflow.

The generalization can be seen to be insightful if the economy – purely analytically – is divided into three sectors: the first producing investment goods, such as machine tools, instruments, and plant equipment, the second supplying intermediate goods, such as minerals, energy, and materials, and the third producing final consumer goods. All sectors are engaged in product and process innovation. However, to the extent that new production methods require qualitatively new investment goods and/or intermediate inputs but no consumer goods, process innovation in the economy originates mainly in the first and second sectors, and particularly in the first. In contrast, product innovation in the consumer sector contributes only to process innovation in the household sector. The basic ideas underlying a new process may of course originate anywhere, in firms as well as in households. However, unless the idea concerns matters like organization or marketing, the R&D burden of implementing the new process usually lies with the development and initial production of new specialized equipment. The subsequent diffusion of the process can also be partly dictated by the supply of that equipment. It may also be worth noting that, while the production of technologically complex products tends to require sophisticated processes and equipment, the reverse is not true; advanced methods may and often are used to produce the same standard goods.

Although the interdependence between the rate of product innovation in the investment goods sector, on the one hand, and the rate of all innovation in the economy as a whole, on the other, is particularly evident in a closed economy, it also has powerful implications for an economy engaged in importing investment goods from the technologically more advanced countries.

Dynamic economies of scale, product cycle, and innovation

Empirical evidence indicates that most new products are minor improvements of existing goods. The methods of producing these goods tend to be well developed already, and hence further improvements in these methods are also minor.

However, this is typically not the case with radically new products, such as the motor car, telephone, and radio at the beginning of the twentieth century, the electronic computer, nuclear reactor, and semiconductor in the 1940s, or the microchip in the 1960s, to give a few extreme examples. The unit cost of producing an important new product is usually very high compared with its use value, but so is the potential for improvement in the performance characteristics of the product and the methods of its manufacture. These improvements may require the

15

invention of new tools and process equipment, much higher standards of purity of the materials used, better quality control instruments, and new labour skills, and these may become available to the producer only gradually in response to the demands emanating from the development work on his new product. In the product's introduction phase the quantities produced are small and the firm itself may also be small, but process and product improvement innovations are fast. As the product becomes cheaper and more attractive, demand for it rises, and with it the scale of production and the size of the firm; this is the growth phase of the 'product cycle'. As soon as specialized equipment for large-scale production is developed and starts to be used, this typically reduces unit costs drastically, setting in a 'virtuous circle': from further reduction in price, to increased demand, to increased scale of output, to still further reduction in price. The producer is seen to benefit from what may be called 'dynamic economies of scale', a combination of three factors: (i) adaptive learning by doing by the work-force and management that reduces cost; (ii) a sequence of product improvements that increase demand for the product at a given price; (iii) positive feedback effects resulting from the increased scale of output on unit cost, innovation efforts in the supplying industries, and the product's price and demand. The growth phase ends and the mature phase begins when the quantity demanded levels off, possibly to decline later on. Two interesting aspects of this phase are the simultaneity of large output of the product, supplied by firms which in the meantime have become large, and low further product innovation as the originally large potential for improvement has become almost exhausted.

The above analysis suggests that the rate of innovation for firms whose fortunes are based on one basic product may be associated with firm size, and this association would mirror the product life cycle. In the introduction phase of the cycle the potential for innovation is vast, but the firm's resources are small. In the mature phase the resources are large, but the innovation potential is small. In the intermediate phase, however, the firm's resources are sizeable and the potential for improvement is high; consequently, the rate of innovation and growth are likely to be highest in that phase. Faced with low productivity of the R&D input and stagnant or declining market demand, large firm(s) in a mature industry may deploy various defensive measures. These might include increasing R&D expenditure, higher advertising effort, developing more distant markets, and seeking to eliminate other producers in the industry. However, the adoption of a diversification strategy, designed to phase out outdated products and deploy more R&D

inputs and other resources to develop and market new products, is often the only effective defence.

The maximum size of the market for a product varies greatly from one product to another. Therefore the meaning of the terms 'medium' and 'large' in the statements above must also vary. In some industries our highly innovative and fast-growing 'medium' firms would be 'large' by the standards of other industries. Consequently, the qualitative relations, implicit in the product cycle hypothesis, between innovation effort and innovation rate and between firm's output growth and firm size need not hold for a group of firms from different industries at any given point in time, even if the relations actually operate for each individual firm from the group over time. However, it is interesting to note that one of the conclusions which Kamien and Schwartz (1975: 3) reach in their extensive survey paper is that 'the bulk of the evidence indicates that, among firms engaged in R&D, relative effort tends to increase with size up to a point and then decline, with middle-size firms devoting the most effort relative to their size', which is precisely what the product cycle hypothesis predicts.

The trigger effect and an illustration of the long-term effects on prices

We may recall that the technology set for a firm comprises all production processes that are known to the firm. These processes may give rise to a large number of alternative techniques of producing the good(s) that the firm wishes to supply. Of these techniques usually only one, chosen presumably for a reason, would actually be operated. Thus, at each moment in time, only a fraction, possibly a very small fraction, of the total number of processes is used, and thus only a small subset of the total technology set is active. The other processes remain dormant, waiting to be used when some of the binding (supply, demand, or financial) constraints change or when prices change. Our discussion also implied that a change in the firm's technology and a technological change are possible without any change in technique.

Following Herbert Simon (1951: 268) and Michio Morishima (1964: 115–22), we shall speak of a 'trigger effect' being present if a new (process or product) innovation in one firm leads to a change in the environment of prices and constraints that induces a substitution of one technique for another in one or more of the other firms. The new innovation in this case must be substantial enough in terms of its price and other economic implications to trigger off a chain reaction involving possibly several firms from

17

different industries and resulting in some of the dormant inventions of the firms in question becoming active. Such a chain reaction usually takes a long time to complete. In the meantime other innovations appear and some of these become new sources that radiate waves of further change. If we looked at the economy as a whole we would observe the arrival of a flow of technological changes in the form of innovations. These can be called the primary changes: they induce another flow of secondary and further changes in which substitution of some techniques for others takes place within the firms' existing technology sets.

The precise nature of long-term trigger effects is, in general, highly innovation-specific, with the one obvious general feature that new consumer goods are incapable of such effects. Some definite results concerning trigger effects can be obtained if we assume, with Morishima (1964), that input–output coefficients are scale independent, each process involves only one output, there are no capital gains, and the rate of interest is fixed. If we simplify further, as William Nordhaus (1969) does, by assuming that there is no fixed capital, there is only one labour input, and the n-sector economy is capable of producing positive net output of all n goods (the economy is 'viable'), then the following two instructive results can be shown to hold:

(i) if in industry i, $i = 1, 2, \ldots, n$, a newly adopted process lowers the cost of producing the ith good at the initial equilibrium prices, then this adoption triggers off a movement of the whole economy to a new equilibrium in which all prices, when expressed in terms of labour's nominal wage as the numeraire, will be lower than the initial prices;

(ii) the price of the ith good will decline proportionately more than the price of any other good.

The equilibrium prices referred to above are the competitive (minimum cost) prices which, in the case considered by Nordhaus, equal both the average and the marginal cost. Since the prices of all goods fall, proposition (i) implies that the new real wage is higher than the initial one, whatever the composition of the goods that the constant nominal wage buys.

Sometimes an assertion is made that, while product innovation tends to stimulate employment, process innovation usually does not. This assertion may be true for particular firms and sectors, but is doubtful for the economy as a whole. A product innovation by a firm may be an effective way of attracting demand for the firm's output at the expense of other firms; it leads to a redistribution of employment to the advantage of the innovating firm rather than to any employment gain at economy level. In contrast, a process innovation – and we presume that it is labour saving – implies a fall in the relative price of the firm's product. That fall would increase the demand for that product, but should that demand be price-elastic this increase would not be enough to stimulate the firm's employment also. In the latter case the other firms are likely to benefit from the ensuing redistribution of employment.

Economic growth and aggregate measures of innovation

Most of the modern mathematical theory of growth began with the advent of major innovative ideas due to Ramsey, von Neumann, and Harrod in the 1930s, and has continued with extensive ramifications and generalizations by a large number of authors in the post-war period. The emphasis of the theory has been largely on the growth effects of capital accumulation under various specifications of the initial production technology and, most importantly, assuming that the subsequent technological change is both given and cost-free. Inevitably, it was found that the growth path of the economy depends to a distressing degree upon the choice of assumptions concerning the innovation rate and bias. The range of specifications of technological change had to be somehow narrowed down, and those models which were shown to produce the growth paths broadly consistent with the observable (so-called) stylized facts have promoted technological change to the role of the main force behind both capital accumulation and output growth. Numerous empirical studies over the last 30 years or so point in the same direction. Despite occasional dissent, it appears that a consensus has emerged that the observed strong variation in the growth rates of outputs among nations and over time is largely the result of differences in the rate of change of the productivities of the non-producible or primary inputs, especially labour inputs. The cumulative effect of this theoretical and empirical work has been to highlight more sharply and widely than ever before how really central is the role, in long-term economic growth, of the activities producing qualitative changes in the economy. Technological changes have assumed the primary role by virtue of their being typically the original impulses which tend to initiate other qualitative changes. By the same token the work has also helped to delineate the very limited usefulness of the (standard) growth theory based on the assumption that these qualitative changes are cost free and exogenously given.

In order to illustrate more explicitly the interactions that take place between technological change, capital accumulation, and

19

output growth, let us consider the case of a closed economy in which all inputs can be classified into the following four categories: labour force L, physical capital (including land) K, human capital H, and technology, or the state of knowledge relevant to production, T. As usual, human capital refers to the skills embodied in the labour force, reflecting its familiarity with and capability of using the available technology, and the latter in turn is embodied in physical capital. Either flows of services of these inputs or the inputs themselves are used, and they would be used in co-operation in a large number of different processes to produce the following flows of output: consumption C (which may include additions to knowledge not useful for the purposes of production), additions to K, additions to H, and additions to T. Aggregating over these processes, we can thus write $Y = F(L, K, H, T)$, where Y represents the flow of aggregate output and L, K, H, and T now denote the flows of services of the respective inputs. The labour and capital services are measured in man-hours and capital-hours. We do not have a convenient measure for the services of H and T, but we associate qualitative changes with a rise in T or H, or both.

Since the inputs are used in conjunction, it is not useful to ask what the contribution of any one input is at any particular moment of time to the total flow of output. If all inputs of any of the input categories were zero, the output would be zero or near zero. In this sense each of the aggregate inputs is essential, and all are responsible for all the output; this point has been emphasized by Matthews (1973) among others.

Therefore we usually consider small changes in outputs, such as those occurring in a 1 year period, and ask what would be the change in output if any particular input changed while other inputs were kept constant. When the changes in inputs are small it is legitimate to presume that the resultant output effects are independent of each other. It is this independence property that enables us to separate out the individual contributions of the small changes in input to the total change in output.

However, this property is not maintained when we consider a process of economic growth over a substantial period of time. The reason is that a change in any one input at the beginning of the period would change the output flow over the period and, given the distribution of output between consumption and additions to inputs, would also change the supply of the other inputs, thus giving rise to further indirect changes in the output flow.

To illustrate this difference between the short and long term more clearly, let us simplify further by writing $H = H(t)$, $T = T(t)$, and $Y = F(L, K, t)$ where t represents time. If we assume constant

20

returns to scale in terms of L and K, the standard growth equation is as follows:

$$g_Y = \pi g_K + (1-\pi)g_L + \lambda \tag{1.1}$$

where g is the growth rate of the variable indicated by the subscript, π is the elasticity of output with respect to capital, and $\lambda = (1/Y)\partial F/\partial t$ is the contribution of the qualitative changes to the output growth. This direct contribution is of course instantaneous, but we shall call it short term. If qualitative changes have been absent in the past, but will be present and maintained at a constant rate λ in the future, then not only g_Y but also g_K would rise, producing an additional indirect contribution to g_Y. What is its size?

The answer depends on the decision concerning the distribution of the direct contribution between investment in physical capital and other uses. Suppose that we keep the share of capital investment in output unchanged. Then a rise of λ in g_Y immediately increases the growth rate of investment by λ and, over time, it also increases g_K by λ, thus producing a feedback effect on the growth rate of output of size $\pi\lambda$. This additional increase in g_Y raises the growth rate of investment, and hence g_K, by a further $\pi\lambda$ and has an additional feedback effect on g_Y of size $\pi^2\lambda$. The appearance of qualitative changes at a constant rate λ is seen to generate what can be called a *growth propagation process*, whose contribution to g_Y will eventually be equal to

$$\lambda + \pi\lambda + \pi^2\lambda + \ldots = \frac{1}{1-\pi}\lambda \tag{1.2}$$

Let v represent the capital to output ratio K/Y. Note that $g_K - g_Y = g_v$. Hence (1.1) can be written in the form

$$g_Y - g_L = qg_v + \alpha \tag{1.3}$$

where

$$q = \frac{\pi}{1-\pi} \qquad \alpha = \frac{\lambda}{1-\pi}$$

The difference between λ and α is very important and should be well understood. In particular it should be noted that α, although referring to a long-term period, is not necessarily the actual contribution of technological change during that period. Since both λ and π, defining α, are current, α is only an approximation, or a forecast, of the true contribution of technological change to growth over any long-term period. This forecast is based on short-term information, and so it will prove correct in the event that the

21

information remains unchanged over time. Moreover, our 'growth propagation' argument assumes that both capital and output grow at a common rate and that both the savings ratio and the employment growth rate are constant. These are the properties of an economy that moves along a balanced growth path. It follows that, while λ is an exact measure of the *direct* or short-term contribution of technological change to output growth, α is a balanced growth forecast or approximation of the total eventual contribution over a longer period of time. (These and other measures of technological change are discussed further in Chapter 8.)

Since net capital investment $K = sY$, and since $K = vY$, it follows that $\dot{K} = \dot{v}Y + v\dot{Y}$ and therefore that

$$s = vg_Y + \dot{v} \tag{1.4}$$

Relations (1.2)–(1.4) are correct whatever the change in v with time. The capital-to-output ratios actually observed appear to have been much more stable in the long term than the corresponding labour-to-output ratios. For those situations at the economy or sectoral level, when it is legitimate to consider the capital-to-output ratio to be approximately constant, relations (1.2) and (1.4) assume a particularly simple form:

$$g_Y = g_L + \alpha \qquad s = vg_Y \tag{1.5}$$

For a number of reasons, of which the use of small data samples is the most important, the econometric estimates of π and λ cannot be and are not reliable. However, in those cases in which the capital-to-output ratios are approximately constant, the long-term contribution of qualitative change to output growth is seen from (1.5) to be (approximately) equal to the growth rate of labour productivity $g_{Y/L}$, which is equal to $g_Y - g_L$ and we can measure this difference directly and more reliably. We also infer from (1.5) a point of central importance: namely, that given the observed small variation in K/Y over long periods of time the wide international diversity recorded in the growth rate of labour productivity is approximately equal to, as well as implied by, the cross-country variation in the rate of qualitative change.

When the labour participation rate is constant, so that the nation's population and man-hours of work change at the same rate, and when the savings rate is also constant, so that consumption and output grow at a common rate, then α is seen from (1.5) to equal the rate of growth of consumption per person. The labour participation rate and the savings rate can and do change with time, but when they are relatively stable, as in fact they tend to be

over substantial periods of time, then given the initial and important assumption of constant returns to scale, it is correct to attribute
approximately all the long-term rise in the average consumption
and income per person in any country to all the past qualitative
changes that have accumulated in that country. Capital accumulation still remains essential as a necessary medium without which
qualitative changes cannot be fully effective in the short run and
cannot take place in the long run.

This intimate interaction between qualitative changes and
capital accumulation makes a fully specified growth model, and
not merely a production function, the appropriate instrument for
measuring the contribution of qualitative changes to growth over a
period of time, i.e. for contrasting the contribution of a change in
the savings ratio – rather than in the growth rate of capital stock –
with the impact of a change in α. This point is important and will
be discussed further in Chapter 8. The reader may also consult
Dan Usher (1980) who, in a chapter entitled significantly 'No
technical change, no growth', gives a careful discussion of the
production function estimates of the contribution of technological
change to economic growth by Solow (1956), Denison (1967), and
Griliches and Jorgenson (1967), as well as of the bias in such
estimates arising from the changing qualities of the inputs and
outputs and from the use of various price indices for aggregation
purposes.

Correction for the changing (static) X-inefficiency

In writing that $Y = F(L, K, t)$ for a total national economy or its
major sector, we mean to capture the grand production
relationship between inputs and outputs on the production possibility frontier, implying full utilization of all the inputs. The
assumption that actually observed inputs and outputs lie on the
production possibility frontier is likely to be unrealistic not only in
the short run – say, during a depression – but also in the long run.
Suppose that u_K and u_L are in fact the utilization rates of K and L
respectively calculated at some 'standard' intensity of work. Hence
$Y = F(u_K K, u_L L; t)$. Instead of (1.1) we now have that $g_Y - g_L =
q(g_v + g_{u,K}) + g_{u,L} + \alpha$, where $g_{u,K}$ and $g_{u,L}$ are the growth rates of
u_K and u_L respectively. If v does not change much over time, the
term qg_v may be small enough to be omitted. Consequently
equations (1.5) become

$$g_Y \approx g_L + \alpha + g_u \qquad s = vg_Y \qquad (1.6)$$

where $g_u = g_{u,K} + g_{u,L}$. These two aggregate growth relations are useful as a starting point in the analysis of long-term growth. Given α and g_u as parameters, the equation should be interpreted as determining any two of the four variables g_Y, s, g_L and v, with the remaining two variables to be regarded as policy instruments.

Chapter two

Inventive activity: distinct characteristics of nature and size

Public good quality of invention and game aspects of the invention/innovation process

So far we have treated the nature of the production process as if there were no differences between conventional production and the inventive activity. To be sure, there are some obvious similarities: both involve inputs and outputs, and the choice of suitable production techniques is largely economically motivated. However, invention and innovation have a number of important distinct characteristics that make invention a qualitatively different product and R&D a fundamentally different activity. We begin by detailing some of the more important of these characteristics and broadly discussing their economic implications.

We can view invention as the production of information, innovation and imitation (diffusion) as its utilization, and inventive activity as a game with nature.

Unlike conventional goods, which are partly or wholly private, invention has the qualities of a purely public good. There are two immediate implications of this characteristic. Firstly, the equivalent of fixed cost in conventional production is the total cost in the production of inventions. It follows that in deciding how widely an invention should be used in the economy, society should be advised to disregard this cost and continue diffusing the invention as long as the marginal cost of diffusion (transmission and assimilation) exceeds the accumulated stream of resultant future economic benefits, suitably discounted. If a social optimum is to be achieved, inventions should be supplied free of charge. However, in a market economy the incentive to produce an invention would then be reduced, and in fact limited only to the expected gain to be derived from the producer's own use of the invention. There are thus two sides to any royalty-generating arrangement such as the patent system: it induces search for inventions and hence reduces

25

their under-supply but, at the same time, it promotes their under-utilization.

The stock of undiscovered ideas can be regarded as a common property resource, very much like the stock of uncaught fish. The over-fishing model of Scott Gordon can then be used to argue for the possibility of *over-investing* – investing more than is socially optimal – in the search for ideas in the situation when access into inventing is completely free and the discoverer is granted full property rights to his ideas (for references see Hirschleifer and Rile 1979). Such a rush to invest may occur when many potential discoverers search for the same inventions or for similar solutions to the same problems. Presumably, some discoveries made by any one of them would then have been made anyway by others.

The implication of the over-fishing model in the R&D context is that, while the social optimum calls for the ideas already discovered to be unrestricted, a right to engaging in the search for an idea should be given only to the lowest-cost inventor, i.e. the one who, if the right were auctioned off, would bid highest for it. In practice, the allocation of R&D resources among different research teams is often intended to follow precisely this principle. However, when time is of essence to the society and when it is uncertain who the lowest-cost inventor is, a certain degree of multiplicity of the R&D effort could well be a socially optimal course to follow.

The under-utilization of an invention might be expected to be lower when the inventing firm alone represents a large share of the potential market for the invention. Therefore the second implication of the public good nature of invention is that increasing returns to scale operate in the use of invention; the return to carrying out R&D will be larger, the larger the firm. This relation would seem to imply that in the absence of both publicly financed R&D and suitable licensing agreements, an economy dominated by large firms would always be more inventive and innovative than an economy which is highly fragmented and competitive. This in turn would imply the presence of a conflict between dynamic efficiency – the rate of output growth given primary inputs – and static allocative efficiency. Schumpeter (1943) gave this implication the status of a major principle which is now usually known as the Schumpeterian hypothesis. Furthermore, this conflict can be expected to be particularly pronounced when the inventions produced are capable of generating large positive externalities. The reason for this is that the gap between the socially optimal and the actual supply of inventions should then be significantly wider under a competitive market structure than under a monopolistic structure. We shall discuss this possibility in some detail in Chapter 4. At this stage it must suffice to say that positive externalities and economies of scale are only two of several important factors influencing the generation and use of inventions, and that there are grounds for expecting situations to occur (e.g. those discussed already on pp. 16–17) in which small and medium rather than large firms would favour or be associated with faster invention and innovation.

The other important characteristic of the inventive/innovation process is that it involves in an essential way uncertainty or even ignorance, where the latter means a lack of knowledge about the densities of the relevant probability distributions. Uncertainty and ignorance are also present in the operation of the conventional sector, but judgements can usually be made about the expected volumes of inputs and outputs since past experience can be used to generate density functions of the probability distributions of the relevant random variables. It also helps that the standard errors for inputs and outputs are fairly small in virtually all conventional economic activities, with the notable exceptions being some types of extraction and agriculture. Because inventive activity involves uncertainty to such a high degree, different R&D teams and organizations may disagree widely as to where to place their 'R&D chips' and when to make their bets (Nelson and Winter 1977a).

What might be thought to be the major implications of this intrinsic characteristic for the conduct of R&D activities? In seeking a preliminary and broad answer to this question it is useful to distinguish between the generation of new inventions and what Nelson and Winter (1977a) call the 'selection environment', which is the assimilation process. The generation of inventions can be further divided between organized technological invention, which denotes all innovations in the economy that occur within R&D organizations as a result of actions by their members, and other non-organized innovations. Sometimes an invention may be made purely by chance, as a by-product of other work, rather than as a result of careful search. However, in perhaps most cases, especially in the organized branch, R&D projects are selected beforehand and the problem is to find the 'best' way or strategy to proceed. Research projects and research strategies can also be considered jointly. When ignorance looms large, the selection of interesting projects cannot be formulated in terms suitable for 'objective' assessment. In this typical case judgements about the probabilities of success must be and are made on the basis of experience and intuitions of the individuals involved. Therefore the most 'promising' projects and strategies that will eventually be chosen must

27

:pend on who the individuals are. However, researchers and their anagers vary significantly, sometimes even quite widely, in terms of talent and knowledge, cultural background, institutions with which they deal, and the organizations in which they work. Consequently, very different 'optimal' choices might be made by different research teams. At the same time, however, some common choice-forming factors are present, and we shall discuss these in Chapters 5, 6 and 7.

The room for variation in the choice of R&D projects and strategies may also be significantly limited by so-called 'technological regimes'. The point is that the basic elements of the car engine, the aeroplane, or the television set can be somewhat modified and improved, but cannot be removed or substantially changed. Technological regimes are usually defined by the initial (basic) discoveries, which are relatively few in any particular branch of economic activity. The endless derivative improvements that follow the initial discovery are further guided by what Rosenberg (1969) calls 'technological imperatives' – things like bottlenecks in connected processes and obvious weak spots in products. They become clear targets for improvement to different researchers at the same time, often worth undertaking under a wide range of particular demand and cost conditions. The research strategies adopted in these cases may not be very dissimilar in terms of the range of inputs and methods employed. The resultant advances would in this case follow what Nelson and Winter (1977a) call 'natural trajectories'. The particular demand and cost conditions, as well as the particular people carrying out the R&D involved, may still affect the choice of particular strategies within such trajectories, giving scope for R&D competition and rivalry within industries (see Chapter 4). A parallel can be drawn between limited product differentiation under monopolistic competition in conventional production and this variation in R&D strategies. More importantly, these variations in market conditions and R&D people may affect the distribution of the R&D effort among the trajectories, with the industries associated with some trajectories expanding faster than those based on other trajectories.

The degree of uncertainty and ignorance is probably much greater in the more basic areas of R&D, those which are fertile ground for major discoveries. Since these discoveries are rarely anticipated, and yet when actually made give rise to new natural trajectories and to corresponding new industries, the detailed input compositions and industrial allocations of the R&D effort are inevitably also subjected to a high degree of uncertainty.

A high degree of uncertainty about the success, or even about the underlying probability distributions of success, may also provide fertile ground for the emergence of 'conventional rules' to guide both the selection of R&D projects and the choice of inventions for application along natural trajectories of a different type. A good example concerns scale effects (Nelson and Winter 1977a). Since in many cases it has turned out that larger-scale processes are cheaper, a convention begins to operate that directs the R&D staff to seek improvements along this route in *all* industries. Similarly, if engineer-designers find, as apparently they did in the nineteenth century (David 1975), that there is often a 'lot of room' for capital-saving improvements of the mechanical operations in capital-intensive rather than labour-intensive processes, this again, when widely noticed, becomes a grand 'rule of thumb' that directs R&D managements and inventors to seek innovation of a particular type. Conceivably, the contemporary convention stressing the need for fast computerization might be a modern example of the same kind. These conventional rules originate from experience and are probably strongly influenced, initially at least, by profitability considerations, actual and potential. However, with time they also tend to develop the quality of a myth which, under uncertainty, may easily become an influencing factor in its own right.

Surges of basic inventions, innovative potentials, and variations in innovation rates

In the previous chapter we have already made use of the distinction between the original few basic, or radical, inventions and innovations and the subsequent sequence of a large number of, usually minor, improvement innovations. The analytical power of this distinction lies in the apparent fact, now reasonably well evidenced by economic historians, that the rise of substantially new branches of economic activity and radical changes in the existing branches can almost always be traced to the discovery, respectively, of substantially new products and radically new processes.

A fairly recent and extensive survey study of these innovations has been undertaken by Gerhard Mensch (1978). The most striking, though not completely unexpected, finding of his study is that the rate of arrival of such basic innovations has varied quite considerably over the past two centuries. Their frequency distribution is seen in Figure 2.1. The time unit chosen is 10 years. A different choice of time unit and initial year, not to mention the inevitably somewhat arbitrary choice of inventions and their discovery dates, would no doubt affect this distribution. However, the fluctuations in Figure 2.1 are so pronounced that it appears

29

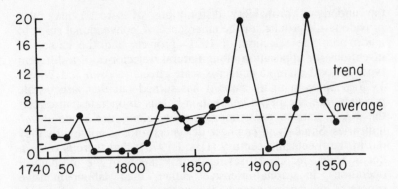

Figure 2.1 The frequency of basic innovations in twenty-two 10 year periods from 1740 to 1960.

Source: Mensch, G. (1974) 'Institutional barriers to the science and technology interaction', in H.F. Davidson *et al.* (eds) *Technology Transfer*, Leiden: Nordhoff, p. 215

safe to accept that basic innovations do tend to come in clusters, with periods of unusually high inventive output being followed by what Mensch calls 'technological stalemates'.

This remarkable characteristic of the inventive activity has the potential of being important as well. This would be so if the surges in basic innovation are shown to be capable of producing, later on, significant fluctuations in the overall innovation rate and in the level of the aggregate economic activity of major countries and the world economy. The output cycles of the Kondriatieff type come immediately to mind. In particular, a natural temptation is to associate the post-war economic upswing with the fast expansion of industries based on the major inventions of the 1930s, and the post-1970 growth slowdown with a partial exhaustion of that growth potential and with the technological stalemate of the 1940s. The usual argument in favour of these two associations, such as that developed by Mensch (1978) himself, does not lend itself easily to empirical testing. The characteristic feature of his theory is the suggestion that firms are particularly keen to explore radical technological ideas in periods of deep economic depression, much as societies adopt radical institutional changes in periods of social distress such as civil wars and revolutions. A similar idea, except that it refers to all innovation at any time by firms in their individual circumstances, is an important part of the Nelson–Winter evolutionary theory (see Chapter 5). However,

Mensch's attractive suggestion is disputed by Clark *et al.* (1981a,b) on the grounds that the case histories of the basic innovations of the 1930s do not mention depression as a factor that 'influenced the decision-making of innovators in a positive way, whether by the initiation of a new development project, by the acceleration of an existing project, by the earlier-than-expected launch of an innovation, or by taking a completed or nearly-completed project "off-the-shelf"' (1981b: 317). Nevertheless Clark *et al.* accept that basic innovations tend to come in bunches but suggest that any such bunching is probably 'related to breakthroughs in fundamental science and technology, to bursts of invention, and to periods of very strong demand (including booms and wars)' (Clark *et al.* 1981b: 3). The ultimate effect of bunching of this type is linked to the presence of the following sequence of casual relations: from basic innovation to overall innovation, to induced investments in fixed capital, to aggregate demand and employment. We may note, however, that some economists, in particular Nicholas Kaldor (1961) and Jacob Schmookler (1966), regard the rate of investment itself as the major factor determining the rate of innovation. In the sequence hypothesis of Mensch the causality is suggested to run in the opposite direction, from innovation to investment, rather as in the theories of the business cycle due to Joseph Schumpeter (1939) and Michal Kalecki (1962). Clearly, whether long-term waves in basic innovation are autonomous and give rise to long-term cycles in output and unemployment or are responses to these cycles hinges on the direction and strength of the relation between innovation and investment about which there do not appear to be generally applicable rules. However, even if only the first relation of the sequence holds, from basic innovation to overall innovation, the innovation surges would still be of major consequence by virtue of their contribution to the variation over time in the growth rates of input productivities and sectoral outputs. In Part II we shall argue that there are also other, possibly more powerful, forces at work – especially in countries outside the technological frontier area – that differentiate the rate of productivity growth over time.

The link between basic innovation and overall innovation hinges in part on the concept of innovational potential. We have already alluded to this in Chapter 1 when outlining the product cycle hypothesis. The basic idea is that the pioneering innovations open up large innovational opportunities within new or renewed industries. These basic innovations are therefore followed by series of significant improvement innovations, the economic effects of which are influenced, on the demand side, by the law of diminishing marginal utility of the product in question and, on the supply side, by eventually diminishing marginal returns to investment in the improvement effort. The presence of such a broadly defined potential, defined by the gap between the ultimate technology or product and the state of the art at any particular time, has been fairly well documented for individual basic innovations. The marginal dimishing returns to investment in improvements need not set in in the initial stage following the invention's first application, which is the period of intensive learning and experimentation. Improvement paths of type (i) in Figure 2.2, the famous logistic or S curves, may in fact be more common than those of type (ii).

We shall suggest, and this is one of the central ideas linking the various strands of arguments and theories in the micro-oriented Part I and the macro-oriented Part II of this book, that technology gaps between the potential and the actual can be identified not

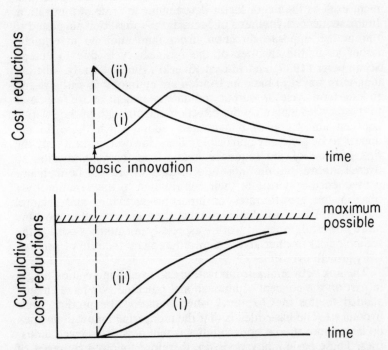

Figure 2.2 Examples of patterns of improvements of an idea following its first application.

only for particular inventions but also at higher levels of aggregation, and that these gaps are useful proxies for the more
fundamental forces which play an important role in explaining the
variation in innovation rate and productivity change among firms,
industries, and countries.

The term 'potential' should be understood to imply the presence
of an implicit or explicit ceiling which limits the cost savings of a
basic process invention or the performance characteristics of a
basic product, even though the sequence of improvements on the
way to that ceiling may be infinite. Such ceilings are imposed by
the laws of nature. The concept of an innovation potential
therefore implies that at some point in time marginal diminishing
returns to improvement would set in, and that thereafter these
returns would be falling to zero with time. In the case of industries
and countries we can speak of relative and absolute potentials. At
any particular time, the relative technological potential for an
industry is the world's 'best' technology at that time. That potential
is therefore changing over time; it is a moving target for the
pursuing firm, industry, or country. The target's movements are in
turn influenced by its own relative potential, as given by the state
of the art in basic sciences at the time, as well as by the ultimate, or
absolute, scientific and technological potential in the field. We thus
have a sequence of relative potentials at a given moment of time,
all bounded by the ultimate potential (when one exists).

Chapter three

Major time trends and cross-sectional tendencies: stylized facts

It is only since about the mid-1950s that technology innovation has become the subject of extensive empirical research and data collection. The possibly dominant part of this work has taken the form of case studies, each based typically on a sample of firms or innovations in a single industry over a short or medium period. Research work of this type has several serious limitations: the sample firms are usually just those which were willing to cooperate; these firms tend to change significantly during the time under study in terms of important characteristics, such as the qualities of workers and management, size, market conditions, products, and methods of production, many of which would nevertheless be assumed to be constant in each firm or the same for different firms; the methods used to obtain the data from firms also vary among studies and the information obtained is often subject to considerable errors. Consequently, the empirical findings that have been reported are neither particularly reliable nor comparable. Some of these serious drawbacks are also shared by the aggregate R&D data for industries and countries. Moreover these data are patchy or missing for the years before the Second World War.

Be that as it may, it is still possible to identify a number of broad patterns which appear to be formed by the accumulated results of these studies and other R&D-related data. These patterns are the stylized facts of innovation. We divide them into time trends and cross-sectional tendencies.

Major time trends

(i) *The role of the industrial laboratory has been increasing and that of the leisure-time inventor declining.*

If in the distant past innovation was largely a by-product of making things, in the course of the twentieth century a specialized

R&D sector has emerged. In recent times this sector has become the main source of innovation, especially in the world's technology frontier area (TFA). After reviewing the evidence for and against generalization (i) Myles Boylan (1977: 110–11) confirms that indeed the bulk of the R&D effort is nowadays performed by large teams of industrial, university, and government laboratories. However, this evidence also indicates that individual inventors still play a major, perhaps dominant, role in those basic and applied research projects where significant originality is crucial, whereas corporate laboratories are more important at the development stage, where above all large financial and human resources are needed.

(ii) *Over the last two to three centuries, the size of the world's specialized R&D sector, in terms of both employment and non-personnel expenditure, has been expanding some three to five times faster than the size of the world's conventional sector.*

An unbalanced growth of the world economy in terms of these two sectors has been a key characteristic of the global technological revolution and industrialization so far. One direct implication of this fact is that a slow-down in the world R&D trend growth rate is inevitable. A fall in the R&D growth in the world's TFA since the early 1970s is perhaps a signal that this trend growth slow-down is already under way.

(iii) *Industrial R&D has been increasingly dependent on the progress in sciences, and the 'science-based' industries have recently become a major growth area.*

Consequently, the spread of scientific results and the trade in machinery and licences are possibly rising in importance as the means of technology transfer, while straightforward product and technology imitation may be less feasible than in the past.

(iv) *The potential for economic gain that inventions create has tended to be noted more quickly and exploited more rapidly with the progress of time in the course of the last two to three centuries.*

This tendency is reflected in the very considerable decline of the average time-lag between invention and first application, as seen in Figures 3.1 and 3.2 with reference to a sample of basic inventions, and in a faster subsequent spread of more recent inventions. Of course, most inventions are minor improvements, and these have always tended to be assimilated much more quickly than major inventions. Consequently, the rise in the average speed of assimilation of *all* new inventions over the three centuries has probably been much less spectacular than that indicated by Figures 3.1 and

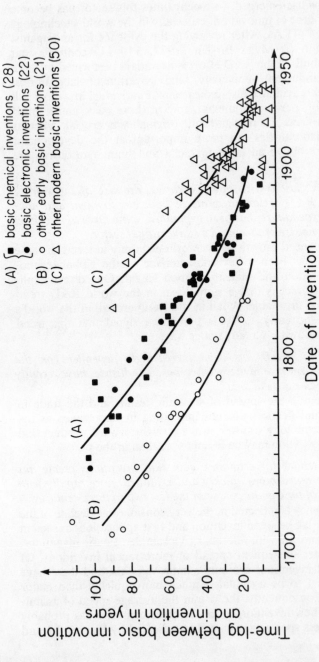

Figure 3.1 The time-lag between the date of basic invention and the invention's first application tends to decline with time. The invention types and their numbers are given above.

Source: Prepared on the basis of data from Mensch, G. (1974) 'Institutional barriers to the science and technology interaction', in H. Davidson *et al.* (eds) *Technology Transfer*, Leiden: Nordhoff, Tables 1–4.

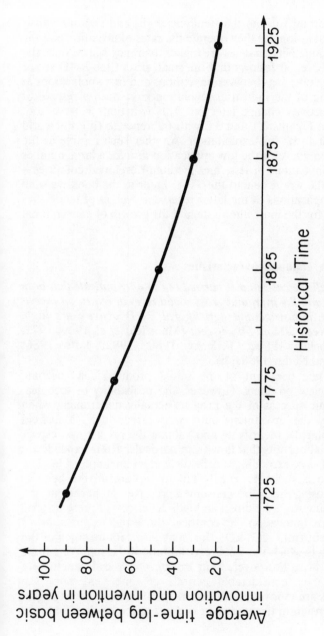

Figure 3.2 The average time-lag between the date of basic invention and the date of the invention's first application declined from 90 years in the period 1701–50 to 67 in 1751–1800, 46 in 1801–50, 32 in 1851–1900 and 20 in 1901–50.

Source: Prepared on the basis of data from Mensch, G. (1974) 'Institutional barriers to the science and technology interaction', in H. Davidson *et al* (eds) *Technology Transfer*, Leiden: Nordhoff

3.2. Nevertheless, it seems fair to assume that this rise has been substantial.

It is important to note that tendencies (ii) and (iv) are transitory, and that as long as they operate the rates of invention and the resultant productivity changes are higher than they will be once the transition is over. It follows that the small size of the R&D sector and the large time-lags between inventions and first applications at the beginning of the industrial revolution have been a reserve of faster productivity change later on. This is a finite reserve and, judging from Figures 3.1 and 3.2 with reference to (iv), the world may be near to its full exploitation. Another such reserve at the world economy level is the low application rate for a large number of modern inventions in less- and medium-developed countries – the part of the world behind the TFA. Later in the book we shall study the implications of the latter reserve as well as of tendencies (ii) and (iv) for the innovation rate and the growth of conventional output.

Major cross-sectional characteristics

(i) *Almost all the case studies carried out to date indicate that both the average and the marginal gross social rates of return on investment in the industrial and agricultural R&D activity are high, usually between 20 and 50 per cent* (Mansfield *et al.* 1969, 1971, 1977; Griliches 1980c; Terlecky 1980, 1982; Jaffe 1986; Bernstein and Nadiri 1988a, b).

Only a few per cent of the initial product ideas become 'successful' new products. However, the probability of technical and economic success of the projects actually undertaken, when weighted by the investment outlays, is rather high (Mansfield 1968: Chapter 3). Identifying and valuing the private and especially the social benefits that flow from particular R&D expenditure is almost always exceedingly difficult and often impossible. The estimated social benefits usually fail to capture all the benefits which the using firms or consumers receive, in particular the benefits arising from the direct and indirect effects of present R&D on all future innovation. In contrast, the estimates of a firm's private benefit from its R&D effort may also capture some of the positive effects which the R&D expenditure of the supplying firms has for the firm. Moreover, both private and social benefits are distributed over considerable periods of time, and since the distributions are innovation- and firm-specific, they add an aggregation dimension in the studies of benefits at the levels of industry

38

and economy. Some of these and other difficulties are reviewed by Boylan (1977: 112–221), Kamien and Schwartz (1975), Griliches (1979), and Link (1987).

To free ourselves from this complex and uncertain detail, let us develop a view of the problem of rates of return which is both economy wide and long term. We noted in Chapter 1, p. 23 that, in the absence of any qualitative changes and assuming constant returns to scale, the net domestic product (NDP) per man-hour of work would remain unchanged over time. However, as a result of qualitative changes that have been produced and absorbed, the NDP per man-hour has been increasing by some 2 per cent in the United States, the main country of the TFA this century. What are the types of expenditures and what share of the NDP has it been necessary for them to have in order to bring about this 2 per cent growth rate? R&D effort of all kinds claims about 3 per cent of NDP. The formal education effort needed to produce new skills takes only a fraction of the total education expenditure. Suppose that the fraction is about a third, which would represent another 3 per cent of the NDP. The third necessary input is net investment in capital assets. Taking 5 for the capital-to-output ratio, the share of the necessary net investment in NDP is 5 × 2 per cent = 10 per cent. Thus a perpetual gain of 2 per cent of the NDP for a given year requires an annual investment of 16 per cent of the NDP of that year, implying that the average social rate of return, economy wide and long term, in the United States has been about ⅛ or 12.5 per cent.

This estimate is admittedly of the back-of-the-envelope type. The precise numbers may be rather different, and they may vary somewhat with the choice of the period. Nevertheless our estimate should be a good indicator of the order of magnitude for the social rate of return on the average R&D and the R&D-related dollar. Our simple estimating method also makes clear the important point that any long-term increase in NDP per man-hour is a joint outcome of the three types of investment – R&D, human capital, and physical capital – and that therefore it is not useful to ask what the separate long-term contribution of any one of these investments is. Yet the composition of the three investments in total spending for the production and absorption of qualitative changes, which is discussed in Chapter 10, is not irrelevant. If the present R&D expenditure were socially suboptimal, this would be indicated by a higher marginal social rate of return from investments in R&D than from investments in fixed capital in the conventional sector.

(ii) *Social marginal rates of return on R&D capital are often much larger than private marginal rates, and the latter in turn are much larger than both the marginal value of physical capital and the interest rate.* A good illustration of these differences is given by the empirical findings of a careful study by Bernstein and Nadiri (1988b), who analysed five high-technology US industries in the period 1958–1981 (Table 3.1). The private return is measured as the variable cost reduction in an industry resulting from its own R&D capital expansion.

The difference between the social and private rates of return measures the extent to which each industry is an R&D capital spill-over source benefiting the other four industries. (Since industries other than the five also benefit, the true social rates are therefore even higher than those reported in Table 3.1). The Bernstein-Nadiri (1988b) study finds that the chemical products industry benefits above all non-electrical machinery, and itself benefits from the R&D of scientific instruments. Non-electrical machinery's R&D spills over on transport equipment, and electrical products' R&D spills over on non-electrical machinery etc. The most significant spill-over is the scientific instruments industry. In another study, Bernstein and Nadiri (1988a) found that output is

Table 3.1   Marginal rates of return in five US industries

Industry	Year	Returns on R&D capital (%)		Returns on non-R&D capital (%)
		Social	Private	
1 Chemical	1961	28.1	19.4	6.7
	1971	21.0	13.2	8.2
	1981	29.1	13.3	13.5
2 Non-electrical	1961	57.7	16.0	7.5
	1971	58.3	26.7	8.5
	1981	45.0	24.0	13.6
3 Electrical	1961	23.5	20.1	7.5
	1971	18.2	15.0	8.4
	1981	30.2	22.4	13.9
4 Transportation equipment	1961	10.5	8.5	7.1
	1971	11.2	9.5	8.6
	1981	16.3	11.9	11.7
5 Scientific instruments	1961	161.5	16.8	8.0
	1971	110.7	17.3	8.3
	1981	128.9	16.1	11.8

Source: Bernstein and Nadiri 1988b: Tables 4 and 5

related negatively to changes in stocks of physical and R&D capital, indicating the presence of adjustment costs. However, the costs associated with R&D are greater than those associated with physical capital. They find that the difference is large enough to explain the observed greater private rates of return on R&D. Another possible reason was given by Schankerman and Pakes (1984), who suggested that the maintenance (or depreciation) costs of R&D capital are high, and therefore gross returns also need to be higher if net returns on R&D capital and physical capital are to be about the same.

In evaluating these findings, we must recall that they refer to technologically leading industries of the world in a period in which R&D effort was already high. Both characteristics tend to reduce marginal rates of return. These rates should therefore be expected to be even higher than those listed in Table 3.1 in countries in which the R&D intensity is yet low and which stand to benefit from the past R&D effort of the developed countries.

(iii) *R&D projects appear to be not only relatively safe and highly profitable but also short term, taking 3 years or less to complete, and to be representing relatively minor advances.*

Judging from UK and US data, about two-thirds of applied R&D projects are expected to be in use within a year of their completion and some 80 per cent within 2 years (Schott 1977; Pakes and Schankerman 1984).

(iv) *The service life for most of the applied innovations is in the range 1–20 years, with the average equal to about 10 years.*

The obsolescence rates are somewhat higher for product innovation than for process innovation (These are again Schott's findings for the United Kingdom.) Higher rates of obsolescence, of some 25 per cent per annum, were found for a sample of US firms by Pakes and Schankerman (1984). Clearly, the rates are bound to vary a great deal among industries.

(v) *The R&D intensity, meaning the R&D expenditure as a proportion of sales, while not evidently dependent on size of firms, tends to be higher in branches whose output expands faster.*

Evidence in support of this tendency's presence in European industry is provided in Figure 3.3. Evidence to the same effect has been produced by Freeman (1982: 24) with reference to UK and US manufacturing activity in the period 1935–58. Indicative econometric evidence for the hypothesis that the association between R&D intensity and growth in sales might be causal, and that the causality tends to run from R&D to growth, has also been

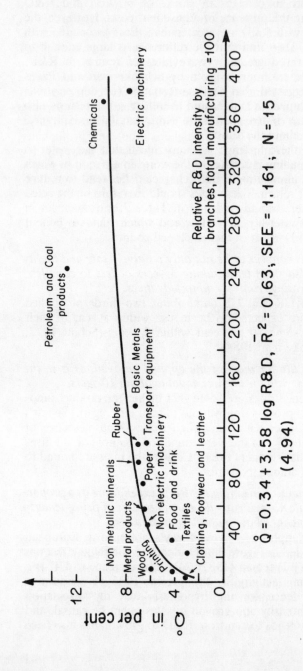

Figure 3.3 The cross-industry relationship between the growth rate of output and R&D intensity. (Unweighted averages of seven West European countries: Belgium, Finland, Federal Republic of Germany, Italy, Norway, Sweden, United Kingdom. The period is from 1958–60 to 1960–70; t-ratio in parenthesis.

$$\overset{o}{Q} = 0.34 + 3.58 \log R\&D, \ \bar{R}^2 = 0.63, \ SEE = 1.161, \ N = 15$$
$$(4.94)$$

Source: Structure and Change in European Industry, UN, 1977, chart 2.2.C, p. 72

reported by Minasian (1962), Comanor (1965), Grabowski (1968), and Leonard (1971). However, in the world of increasing and mutual interactions between innovation and demand, the direction of causality cannot be one way only, and this calls for care in interpreting all the empirical evidence above. While present R&D may stimulate future sales, the possibility is also that higher sales would induce firms to increase R&D effort further and may even increase future R&D intensity. In some cases, higher incomes of the population or expanded trade with other nations and larger scale of production may lead to higher output and profitability, and only *later* to greater R&D. In many other cases, however, new profit opportunities and new fast-expanding markets are created as a result of past innovation. The tendency for the present R&D to influence future profitability and, at the same time, to be influenced by past profitability was indeed confirmed by Ben Branch (1974), who applied distributed lag analysis for data from 111 US firms covering seven industries in the period 1950–65. The short-run causality may thus flow in both directions, with the result that over longer periods of time the trend growth rates in output and productivity tend to be associated with the accumulated research effort; the output growth and R&D effort are both particularly high in those pockets of economic activity where demand and innovative potentials also happen to be high. These varying potentials would then differentiate the industrial branches shown in Figure 3.3 and be a major cause of the positive relationship between output growth and R&D intensity.

(vi) *R&D activity in the world's TFA is characterized by an exceptionally high concentration in relatively few industries and in several dozen R&D programmes. This tendency appears to apply especially to state-financed R&D.*

In 1970 the largest 100 programmes in the United States claimed about 80 per cent of the total US R&D expenditure (Freeman 1982: 133). These programmes were mainly performed in large firms with more than 5,000 employees. However, the research intensity within industries varies widely among firms of similar size, so much so that there is no well-established relation to be found between average research intensity and size. Freeman's own study of innovations in the United Kingdom from 1945–80 found that firms with less than 200 employees contributed about 12 per cent of the total number of innovations while accounting for 19 per cent of net output and 22 per cent of total employment (Freeman 1982: 139). Moreover, 'the vast majority of small firms (probably over 95 per cent) do not perform any specialized R&D

programmes' (Freeman 1974: 205). However, in some industries – in particular where the cost of entry and the capital intensity were low – the R&D intensity of small firms was found to be above average.

The dominant innovation stimulus: technology-push versus demand-pull hypothesis

For innovation to occur and be successful there must be both technological opportunity and commercial viability. Therefore it is not useful to ask what the separate contributions to successful innovation of each of the two factors are. However, in some cases invention is sought for and subsequent innovation undertaken largely in response to the appearance of a need, while in other cases the need is apparent only after the invention takes place. A distinct change in the innovative effort by the US car makers in the late 1970s, following opposite shifts in the demand functions for large and small cars, is an example of a demand-pull innovation process. In contrast, the introduction of a substantially new product is typically followed by a search to find new applications and improvements with the aim of shifting the product demand function. In this case the initial push to innovate comes from the recognition of a technological opportunity, although no doubt the innovating firm hopes that the subsequent search will prove the venture to be commercially viable.

For both the theorist of innovation and the policy-maker it is of considerable interest to know the relative weight of the two stimuli. This was in fact a major aspect investigated by a number of empirical studies, some of them fairly large scale. Summarizing the findings of eight such studies, Joel Goldhar writes that 'all tend to agree ... that between $^2/_3$ and $^3/_4$ of all successful innovations were stimulated by information about a market need (potential demand recognition)' (Goldhar 1974: 35). These findings have come to be widely interpreted as evidence that the dominant influence upon the innovation process is that of market demand. However, this interpretation was challenged by Mowery and Rosenberg (1979: 102–53). They re-examined the empirical and analytical content of eleven of these empirical studies, and found a tendency in them to regard as an innovation stimulus on the demand side not only actual shifts in demand but also the recognition by firms of potential demand. These authors note that in a competitive environment the firms may also justifiably regard the recognition of potential demand as the dominant stimulus in those situations when the initial stimulus to innovate came in fact from the supply side rather than the market-place. Consequently they conclude that 'the primacy of market demand forces within the innovation process is simply not demonstrated' (Mowery and Rosenberg 1979: 739).

This conclusion can be made stronger in the context of innovation and economic growth in the long term. In this context the opportunities presented by technological advance and by shifts in demand are no longer two sharply different types of innovation stimuli, but two sides of the same sequence of changes whereby present shifts in demand are induced by past innovation (rather than by autonomous changes in consumer needs). From that long-term perspective the ultimate innovation forces must be primarily on the supply side. The consumer plays an important role in the short run, even though a largely passive role in the long run, of a selector from among the many alternatives which the supply side is offering. Consumers influence the types of television sets, radios, or computers, but do not influence the invention of these goods because they cannot. This view is consistent with the broad affinity of the types of goods consumed by societies of similar wealth whose cultures are widely separated and different.

The interfirm variation in R&D expenditure: Mansfield's model

Firms which have identified a greater volume of innovation opportunities that promise to yield high net benefits could be expected to spend more on R&D than other firms of similar size in the same industry. The purpose of Mansfield's model (Mansfield 1968: Chapter 3) is to find how much of the interfirm variation in R&D expenditure can be attributed to the variation in firm size and how much to the variation in R&D opportunities. There is also a practical consideration on Mansfield's mind, namely that his model lends itself to empirical testing.

An important distinction is made between the actual and the desired R&D expenditure, denoted R and R^* respectively. In each year the firm is assumed to adjust R from the previous year by moving a fraction θ towards the desired level. Hence

$$R(t) = R(t-1) + \theta(t)\{R^*(t) - R(t-1)\} + z(t) \qquad (3.1)$$

where z is a random term with zero expected value. $R^*(t)$ and $\theta(t)$ must be explained, and R^* is explained as follows.

Suppose that a list is drawn up of all R&D projects which are submitted for consideration to the firm's managers. Let M represent the total cost of these projects, only some of which will be

45

found desirable. Consider the probability distribution of this proposed R&D expenditure M with respect to the estimated (expected) rate of return, denoted ρ. The upper tail of this distribution is assumed to be of the Pareto type; the probability that ρ exceeds an arbitrary rate x is thus Cx^{-a}, where C and α are the distribution parameters. Let ρ^* represent the minimum acceptable rate of return from the successful R&D projects; only the projects yielding ρ^* or more will be considered desirable. Let $\bar{\rho}$ be the average expected rate of return from the desirable projects, and let r be the minimum acceptable rate of return in the industry from risk-free investment projects. The rate r, representing the opportunity rate of return, is assumed to be known. Of course $r < \rho^* < \bar{\rho}$. Denote $\Pr(\rho > r)$ by h; thus h is the fraction of M which corresponds to projects yielding more than r. Since $\Pr(\rho > x) = Cx^{-a}$, it follows that $Cr^{-a} = h$. Hence $C = hr^a$. Substituting ρ^* for x and hr^a for C gives $\Pr(\rho > \rho^*) = h(\rho^*/r)^{-a}$, which is the fraction of M that it is desirable to spend. Hence

$$h\left(\frac{\rho^*}{r}\right)^{-a} = \frac{R^*}{M} \tag{3.2}$$

If economies of scale in carrying out R&D operate, the investments hM should be greater in larger firms. Mansfield assumes that h is scale independent, but that $M = mS^{a_1}$, $a_1 > 0$. There is little justification for either of the two assumptions, but at least they imply that hM is scale dependent. Noting that for Pareto-type distributions $\alpha = \bar{\rho}/(\bar{\rho} - \rho^*)$, we have, for firm i and year t,

$$\log R_i^*(t) = a_0 + a_1 \log S_i(t) + a_2 \frac{\bar{\rho}_i(t)}{\rho^* - \bar{\rho}_i(t)} + u(t) \tag{3.3}$$

where $a_0 = \log(hm)$, $a_2 = \log(\rho^*/r)$ and u is an error term. Mansfield was able to obtain the data on R^*, S, ρ, and ρ^* for a number of firms. These data enable him to estimate the important parameters a_1 and a_2, and hence also the ratio ρ^*/r. Before discussing these estimates, we may note from (3.1) that, in order to explain the actual R&D expenditure R, an equation for $\theta(t)$ must still be developed. Mansfield assumes that θ is higher when the relative distance between the actual and the desired level of expenditure is smaller and when the ratio π of the firm's profits to its actual R&D expenditure is greater. Taking linear specification, we find that

$$\theta_i(t) = a_3 + a_4 \, \frac{R_i^*(t) - R_i(t-1)}{R_i(t-1)} + a_5 \pi_i(t-1) + \tilde{u}(t)$$

(3.4)

where π is expressed in terms of the average value of this profits/ R&D ratio in the industry to which the firm concerned belongs.

The model is thus completed. It is seen to involve none of the optimality conditions which are implied, for instance, by the principle of profit maximization, and which underly the models to be discussed in the next chapter. Equations (3.1) and (3.4) rest on the notion that firms behave in an adaptive manner, with profit considerations influencing rather than deciding their choice of R&D. In particular, the firm's *ex ante* estimate of the profit rate $\bar{\rho}$ is expected to capture all the future implications for profits of the R&D projects that the firm considers taking up. The rate $\bar{\rho}$ is thus a catch-all variable; it reflects the firm's innovation opportunities as well as the special qualities of its R&D inputs and expected demand conditions. Mansfield's model makes good sense when intra-industry differences in all these characteristics are wide and yet cannot be readily explained by reference to measurable economic variables.

Despite the reported poor quality of the data and a sample of only fifteen observations in the US chemical and petroleum industry, Mansfield's estimates of the coefficients $a_2 \ldots a_5$ are statistically highly significant (high ratio of the estimates to their standard errors), indicating that profitability and size are in fact the important variables affecting R&D. The estimated value of 0.75 for a_2 in (3.3) implies that, for the sample in question, the minimum expected rate of return from risky R&D projects was about twice the minimum rate of return from risk-free investment projects. The value for a_1 was found to be 0.89, indicating that larger firms in the sample tended to spend a somewhat lesser proportion of their sales on R&D than did smaller firms. This result need not represent evidence against the presence of increasing returns to scale from innovating. One reason is that larger firms might have produced a wider range of product types than did smaller firms. There is also the possibility that the productivity of the R&D effort declines with the size of the firm's R&D activity – at least for sizes above a certain minimum threshold – inducing some firms to increase that effort less fast than sales.

Chapter four

Market structure, rivalry, and innovation

The central concern of the standard neoclassical theory of the firm is with the choice of profit-maximizing production processes and quantities of inputs and outputs under different assumptions of market structure, time-scale, behavioural perceptions, quality of information, etc. For our purposes the theory needs extending to allow for research inputs and a change in the technology set. Considerable work in building up such a generalized theory has been done recently by a number of authors. This work has been much inspired by Schumpeter's (1942) ideas about the central role of innovation in modern capitalist economies and the roles of the entrepreneur and market structure in the innovation process. Business history provides strong evidence in support of the major importance of 'entrepreneurial talent'. However, for economics theorists this concept is elusive, difficult to define, and difficult to measure independently, i.e. in ways other than in terms of economic results. The academic literature has therefore focused on Schumpeter's argument that seller concentration enhances R&D and innovation. According to this argument, a large-scale monopoly firm is to be preferred to many small-scale competitive firms on two grounds. First, the 'monopoly firm' will have greater demand for innovation because its large size increases the ability to profit from any innovation. Second, the monopoly firm will generate a greater supply of innovations because 'there are advantages which, though not strictly unattainable on the competitive level of enterprise, are as a matter of fact secured only on the monopoly level' (Schumpeter 1942: 101). Fisher and Temin (1973, 1979) interpret this to mean one or all of the following: (i) a larger R&D staff can operate more efficiently than a small one (because, for instance, it allows room for more specialized personnel); (ii) an R&D staff of a given size operates more efficiently in larger firms (it allows room for more diversified activities, lowering the risk of producing knowledge it cannot use); (iii) larger firms can buy more R&D inputs at a given expenditure (e.g. they can borrow finance at less cost).

These propositions seem reasonable enough. Three important implications follow from them: (1) research intensity and inventiveness should be related positively with firm's size; (2) industrial market structures are themselves influenced by past and current innovation, i.e. these structures are endogenous; (3) rivalry and 'creative destruction' lead industries to have less competitive market structures, and with time they may become dominated by single firms.

In a number of recent theoretical papers attempts have been made to formalize the Schumpeterian interactions between innovation, demand, and market structure. Among these contributions are those of Nordhaus (1969), Nelson and Winter (1977b, 1978, 1982a,b), Levin (1978), Loury (1979), Dasgupta and Stiglitz (1980a,b), Futia (1980), and Lee and Wilde (1980). The evolutionary approach taken by Nelson and Winter is distinct in its attempt to model the dynamics of the interaction; we shall discuss it in the next chapter. All the other models are essentially static and highly stylized, concentrating on the end or equilibrium outcome of the interaction. Those of Nordhaus (1969) and of Dasgupta and Stiglitz (1980a,b) are good representations of the group and we shall discuss them in some detail. (A comprehensive treatment of the subject is given by Kamien and Schwartz (1982) and Baldwin and Scott (1987).) These two models will be seen to display Schumpeterian properties. However, it is apparent from the surveys by Kamien and Schwartz (1975) and Scherer (1980) that scale economies, if any, in inventive activity are weak. Neither is there any strong evidence to indicate that the sector of small-scale enterprises is disappearing. Moreover, the socially owned industries in Eastern Europe and the USSR are proving less innovative than their competitive equivalents in the capitalist West, contrary to the expectations of modern theory (e.g. Arrow 1962; Nordhaus 1969; Dasgupta and Stiglitz 1980a,b) and Schumpeter himself. Levin and Reiss (1984) do find some evidence in support of the Schumpeterian simultaneity (of market structure and inventive activity), but they also emphasize the importance of factors, such as the richness of technological opportunities, incentive systems, and managerial qualities which are largely independent of scale. Scale economies in inventive activity may well exist, but it seems that the combined effect of the other factors dominates the innovative process. The models we shall discuss in this chapter focus on scale economies and could therefore be potentially misleading. At the same time, they do inform about important aspects such as the interaction between industry-wide demand and innovation or between one firm's R&D and the R&D effort of other firms.

Market structure, R&D expenditure, and innovation: the Nordhaus model

The firm that Nordhaus (1969) considers has two departments, one producing a conventional commodity and the other cost-reducing innovations. The model is in two versions. The simpler version assumes a perfectly competitive environment so that all firms are price-takers in output and input markets. It also assumes that owing to, for instance, prohibitive transfer costs, no firm may benefit from research conducted by other firms; there are no externalities, no spill-over effect. These two assumptions are relaxed in the second version, in which both the commodity price and the number of firms are endogenously determined, and in which the spill-over effect is present. The competitive solution obtained in the second version is also compared with the optimal solution for the socially managed industry. The comparison implies the classical result (Arrow 1962), namely that market economy typically invests in research significantly less than is socially optimal. expenditure which has been made over the τ period preceding $t =$

Perfectly competitive industry with no spill-over effect

The model is purely static in the sense that only one time period of length τ is considered within which all variables remain unchanged. The volume Q of output flow of the single commodity produced is given by a production function $TF(K,L)$, where T is the level of technology, or efficiency index, and K and L are the services of capital and labour respectively. At time $t = 0$, which is the beginning of the period in question, the firm has the possibility of paying R and obtaining the level of technology $T = T(R)$. Alternatively, the investment R can be thought of as the research expenditure which has been made over the τ-period preceding $t = 0$ and discounted at $t = 0$. The value of net profits generated in the period $(0, \tau)$ and discounted at time zero is then

$$V = \int_0^\tau \{pT(R)F(K, L) - qK - wL\} \exp(-rt) \, dt - R$$

(4.1)

where p is the commodity price, q is the rental on capital, w is the wage rate, and r represents the constant discount rate. The firm has to choose the inputs R, K, and L so as to maximize V. The production function $T(R)F(K,L)$ is assumed to display suitable neoclassical properties to ensure that a positive solution exists. Since prices and quantities are presumed to be constant within the τ-period, we can integrate (4.1) to obtain

$$V = \{pT(R)F(K, L) - qK - wL\}\psi(r) - R$$

where $\psi(r) = \{1 - \exp(-r\tau)\}/r$. Applying the first-order necessary condition for an optimum gives the usual result that the marginal value product of a marginal input equals the cost of the marginal input. In particular, $pT'(R)F(K, L)\psi(r) = 1$, where $T'(R) = dT/dR$, or

$$T'(R) = \frac{r}{1 - \exp(-r\tau)} \frac{T}{pQ} \tag{4.2}$$

If returns to research expenditure are diminishing, so that a lower $T'(R)$ is associated with a higher R, then it follows from (4.2) that the firm's optimal R is related positively to the firm's size, reflecting the proportionality of cost savings to output, and negatively to the interest rate. The latter association mirrors the fact that a higher discount rate reduces the present value of future marginal benefits.

Example. Assume that $Q = R^\alpha K^\beta L^\gamma$, where $\alpha, \beta, \gamma > 0$ and $\alpha + \beta + \gamma = 1$. In this case routine calculation shows that the firm's choice of input quantities is as follows:

$$R = \alpha pQ\psi(r) \qquad qK = \beta pQ \qquad wL = \gamma pQ \tag{4.3}$$

The share of R in the total revenue discounted at $t = 0$ is thus equal to α, a constant.

The specification of the technology function in the example above implies zero output when R is zero. It is clearly more realistic to assume that $T(R) = T_0 + aR^\alpha$, where $T_0 > 0$. In this case we have that

$$\frac{R}{pQ\psi} = \alpha \frac{aR^\alpha}{T_0 + aR^\alpha} \tag{4.4}$$

In response to changes in r and Q, both the optimal share and the level of R&D expenditure move in the same direction and this time α is only the share's upper bound.

Oligopoly with free entry and a spill-over effect

This is a similarly static model in which, however, the commodity price and the number of firms are endogenous. Moreover, it is recognized that each firm may benefit from the research conducted by other firms in the industry. The technology function for the jth firm has the convenient additive form

$$T^j = \sum_{i=1}^{N} \lambda_{ij} H(R_i) \qquad 0 < \lambda_{ij} < 1 \qquad (4.5)$$

where N is the number of firms. Firms are profit maximizing and behave non-cooperatively; each chooses its own output rate, the R&D expenditure, and the quantities of other inputs. There are no barriers to entry into the industry, so that after a sufficiently long time the number of firms in equilibrium is such that, while the last entrant still makes non-negative profits, any additional entrant would suffer losses or, at best, make zero pure profits. The analysis is restricted to the perfectly symmetric case in which firms in the industry behave identically and, indeed, are identical. Accordingly, λ_{ij} is specified to equal a common λ for all $i \neq j$ and to be unity for $i = j$, where λ is in general a function of N. Nordhaus assumes that the input–output coefficients in the production department are fixed, so that

$$F(K_j, L_j) = \min(aK_j, L_j) \qquad L_j \leqslant \bar{L}_j \qquad (4.6)$$

but

$$F(K_j, L_j) = F(K_j, \bar{L}_j) \qquad L > \bar{L}_j$$

Returns to scale in terms of K and L are thus constant as long as L_j does not exceed $\bar{L}_j$, but at that point diminishing returns set in dramatically. He also assumes that

$$H(R_j) = R_j^\alpha \qquad 0 < \alpha < 1 \qquad (4.7)$$

(Again a better specification would be $H(R_j) = T_0 + R_j^\alpha$.) Since firms are identical, $\bar{L}_j = \bar{L}$ and $R_j = R$ for all j. When (4.5)–(4.7) are taken into account, the output of each firm is

$$Q = \mu R^\alpha \bar{L} \qquad (4.8)$$

where $\mu = (N - 1)\lambda(N) + 1$. We note that returns to scale are increasing in terms of all three inputs when L is in the range $[0, \bar{L}]$, and this, given the assumption that demand is independent of price, pushes each firm to employ $\bar{L}$ workers. The capital stock employed is of course $\bar{L}/a$.

The coefficient μ in (4.8) represents the ratio of social to private marginal products. The reason is as follows. If any individual firm increases R by ΔR, its private marginal benefit will be $p\Delta Q = p\alpha R^{\alpha-1}\Delta R = p\Delta T\bar{L}$. However, the extra R&D investment benefits each of the other firms as well by an amount equal to $p\lambda \Delta T\bar{L}$. The total industrial benefit is thus

$$\{(N-1)\lambda + 1\}p\Delta T\bar{L}$$

which is precisely μ times the firm's private benefit.

In this formulation μ is seen to depend on the number of firms. However, the nature of this dependence is not clear, and Nordhaus eventually treats μ as a given (constant) parameter, specific for a particular industry. Since the optimal scale of operation in terms of conventional inputs K and L is predetermined, it remains to choose R with a view to

$$\text{maximizing}\left\{\left(pQ(R) - w\bar{L} - \frac{qL}{a}\right)\psi(r) - R\right\} \qquad (4.9)$$

i.e. maximizing net profits to be had during the $(0,\tau)$ period, discounted at $t = 0$.

Applying the first-order condition for a maximum gives

$$p\alpha R^{a-1}\bar{L}\psi(r) = 1 \qquad (4.10)$$

The zero-profit condition for a long-term equilibrium under free entry implies

$$\left\{p\mu R^{a}\bar{L} - \left(w + \frac{q}{a}\right)\bar{L}\right\}\psi(r) = R \qquad (4.11)$$

Market demand for the commodity is assumed to be fixed at a level $\bar{X}$. Hence,

$$NQ(\text{R}) = \bar{X} \qquad (4.12)$$

An extension to the general case of a price-dependent demand is straightforward except that, in this case, firms may employ less than $\bar{L}$ workers and so we would have one more variable to determine. As it is, equations (4.10) and (4.11) give the equilibrium price and the optimal level of R&D expenditure, while (4.12) determines the number of firms. (If the number $\bar{X}/Q(r)$ in (4.12) is not an integer, N can be taken to be the largest integer not exceeding that number. In what follows we assume that (4.12) holds to a sufficiently good approximation and that input prices are given.) Routine calculation gives the following solution for R, N, and the cost C of supplying $\bar{X}$:

$$R^o = \frac{a}{\mu - a} \, \bar{c}\psi(r)\bar{L}$$

$$N^o = \left(\frac{\mu - a}{a}\right)^a \frac{\bar{X}}{\mu \bar{L}^{1+a} \bar{c}^a \psi^a} \qquad (4.13)$$

$$C^o = \{\bar{c}\bar{L}\psi(r) + R^o\}N^o$$

where $\bar{c}$ represents $w + q/a$, and the superscript o indicates that this is a competitive solution under the oligopolistic market structure.

We define the ratio R/pQ as research intensity, $1/N$ as an index of concentration, $(p - c)/c$ as a measure of monopoly power, and c, which is the unit 'prime' cost of production, as a measure of technical efficiency. (Given the static nature of the model, no distinction can be made between static and dynamic efficiency.) How would these four important characteristics of an industry respond to a change in μ?

The total 'prime' cost of producing Q is $\bar{c}L$ and, by definition of c, this cost is equal to cQ. Therefore, taking (4.8) into account,

$$c = \frac{\bar{c}}{\mu} R^{-a} \qquad (4.14)$$

From (4.13) it follows that when the spill-over effect is greater the level of research expenditure by each individual firm is lower. However, external economies outweigh the adverse effect of lower research, and the firm's level of technology, equal to $\mu(R^o)^a$, rises with μ. From (4.14) we note that the unit cost is related inversely to the level of technology, and hence it falls as μ rises. So does the number of firms N^o, the revenue p^oQ^o, and the ratio R^o/p^oQ^o. (The last two properties are implied by (4.11).) The ratio p/c is equal to $\mu/(\mu - a)$, and so it falls with μ, although insignificantly for $a \ll 1$.

In conclusion, the Nordhaus model implies that, given the zero price elasticity of demand, an increase in the size of the spill-over effect would raise the degree of concentration, reduce research intensity, increase the technical efficiency of the industry, and reduce the mark-up over production cost. Thus, all other things being equal, the degree of concentration and technical efficiency would be positively related in an inter-industrial comparison with μ varying. The monopoly power and research intensity would also be related. These are, in fact, Schumpeterian relations. When a, instead of μ, is the parameter that varies strongly among industries, the model again implies the presence of a positive association between technical efficiency and monopoly power.

The socially managed industry

We continue to maintain all the symmetry and other assumptions of the model above. However, instead of maximizing profits, the overall industrial manager now seeks the number of firms, the research effort and other inputs for each firm so that, given input prices, the combined R&D expenditure and production costs of supplying $\bar{X}$ are at a minimum. The symmetry assumptions ensure that all firms would again be identical. On the assumption that μ is independent of N, routine calculation gives the following solution, here expressed in terms of the oligopolistic solution:

$$R^s = \frac{\mu - \alpha}{1 - \alpha} R^o$$

$$N^s = \left(\frac{1 - \alpha}{\mu - \alpha} \right)^\alpha N^o \qquad (4.15)$$

$$C^s = \frac{1}{\mu} \left(\frac{\mu - \alpha}{1 - \alpha} \right)^{1-\alpha} C^o$$

It is seen that the two solutions coincide only when $\mu = 1$. Guided by the empirical work of Minasian (1962) and Mansfield (1968) for US manufacturing and Griliches (1964) for US agriculture, Nordhaus suggests that 0.1 is a reasonable estimate of α for the entire US economy. We know very little about μ, but judging from Table 3.1 it must vary widely from industry to industry. In any case, it follows from (4.15) that, for $\alpha = 0.1$, N^s and C^s would be close to N^o and C^o respectively for any value of μ from a wide range, say 1–10. However, the third ratio

$$R^s/R^o \approx \mu \qquad (4.16)$$

and so it is strongly dependent on μ. To illustrate the implications let us take, following Nordhaus, $\mu = 2.5$ and $\alpha = 0.1$, so that this ratio equals 0.375, indicating that the level of research under oligopoly might be significantly less than is socially optimal. However, the welfare loss, as measured by the ratio $(C^s - C^o)/C^s$, is 3.5 per cent. This may seem to be low. Given the static nature of the model, the welfare loss should conform with the R&D investment, which is also of the order of 3 per cent of costs. The rate of return from the extra R&D investment, $R^s - R^o$, would therefore be very high. In other words, the ratio $-(C^s - C^o)/C^o$ should be seen as an additional rate of innovation, rather than a mere once-and-for-all fall in costs, which the additional R&D effort is bringing about.

The amount of subsidy required to bring the R&D expenditure to its socially optimal level is thus $R^s - R^o$, or a fraction $1 - R^o/R^s$ of the total (optimal) research costs. On applying (4.15) we can re-express this fraction as $(\mu - 1)/(\mu - \alpha)$. Hence, approximately,

$$\frac{\text{subsidy}}{\text{optimal R\&D expenditure}} = \frac{\mu - 1}{\mu} \qquad (4.17)$$

If the government succeeds in making firms share their innovations more freely, so that μ rises, then, according to (4.17), the share of the government's subsidy should also increase. Incidentally, at $\mu = 2.5$ and $\alpha = 0.1$, this subsidy for US industry should represent 62.5 per cent of the total R&D expenditure, and in the 1970s the actual subsidy happened to be approximately this size. It is tempting to think that this is not a pure coincidence but an indication that the actual μ for the US economy might very well be not significantly different from 2.5. However, the size of the subsidy tends to be particularly high in defence-oriented industries, presumably for other than purely economic considerations.

The method of modelling external economies and the above comparison of the oligopolistic and the socially optimal solutions are perhaps the most interesting aspects of Nordhaus's analysis. To be sure, he makes a number of strong simplifying assumptions. Three of them appear to be particularly important: (i) the profit-maximizing behaviour; (ii) the demand function is totally price inelastic; (iii) the specific form of the technology function (equations (4.5) and (4.7)). Leaving (i) aside for the moment the analysis can easily be extended to cover a wider class of demand functions. However, the quality of the third assumption is more difficult to judge, although perhaps a logistic-type function could be more realistic.

The firms are also assumed to be able to hire at will the quantities of all the inputs they desire, including the high-quality research input, and to assimilate the new cost-reducing processes immediately. Since, according to Mansfield (1980), both the average and the marginal private rates of return on R&D projects appear to be comparatively very high, usually in the range 20–50 per cent, it is legitimate to ask why the R&D expenditure is not higher, perhaps much higher, than it actually is. The above analysis is not capable of answering this question. There seem to be three possibilities. One is that the employment of the research input is limited on the supply side, e.g. by the number of suitably qualified personnel. Another possibility is that the constraining factors are those underlying what is often referred to as the absorptive capability: things like the managerial expertise, the time needed to re-train the work-force, the specialized capital investment required to adopt new inventions, and so forth. Finally, there is the possibility that firms have poor knowledge about the rates of return on R&D projects and, more generally, do not strive to discover and make full use of the profit opportunities that are actually present. While we shall have more to say about this third possibility in the latter part of this chapter and in Chapter 6, its actual significance will remain an open question. At this stage it may be noted that the model's symmetry assumptions imply that all firms are alike. Consequently, while the model can be used to interpret inter-industrial differences in factors such as the rate of inventive activity, the degree of concentration, and the degree of monopoly, it is obviously unsuitable for interpreting in each industry the inter-firm variation in the size of the R&D expenditure, the expenditure's share in the revenue, technical efficiency, output growth, etc. Variations of this type are interpreted by theories of technological diffusion. The key starting point of those theories is rejection of the following two neoclassical premises: each firm always has complete information about the production processes which each other firm in the industry commands, and it can act on it instantly.

Innovation, demand, and market structure: the Dasgupta–Stiglitz model

We have noted in the previous section that an important aspect of the Nordhaus model is the simplifying assumption of an infinitely inelastic demand function. This assumption is relaxed in the Dasgupta–Stiglitz model (Dasgupta and Stiglitz 1980a,b). Consequently, the price elasticity of demand comes into play as another (exogenously given) key parameter of the environment in which firms operate, taking decisions that affect both innovation rate and market structure at the same time. However, the spill-over effect that plays such an important role in Nordhaus's model is assumed away: the spill-over parameter λ is thus presumed to be zero, and μ is identically equal to unity.

Both models share the important assumption of constant unit cost of conventional inputs, so that the total unit cost continues to be given by (4.14), thus displaying economies of scale if the research input is also considered. The presence of such economies represents an incentive for some firms to fuse their resources and earn extra profits before other firms do likewise. In the absence of any legal constraint, the industry would normally collapse into a single firm. One way of preventing this happening, in a theoretical model, is by limiting the range in which the scale economies operate, and this is the assumption which Nordhaus makes. Another way is to assume that all firms always behave identically. This behavioural assumption would preclude non-symmetrical fusions and so remove the incentive to fuse in the first place. Its realism is not clear, and is probably suspect. Dasgupta and Stiglitz adopt this important assumption in their model. As a result the difference between the two models is so substantial that, as we shall see, it renders their respective implications largely incomparable.

Still common to both models are a number of key assumptions: product innovation and any non-price competition are absent, the industrial demand curve is non-shifting, process innovation is divorced from investment in plant and machinery, firms within the industry are identical in every respect, a once-and-for-all R&D investment buys a level of unit cost, rather than a sequence of investment decisions buying a sequence of changes in unit cost, changes in factor prices are absent, and firms know their R&D technology, strive to maximize profits, and can adjust the employment of inputs to their optimal levels at will. These are, of course, extremely unrealistic assumptions. Yet the model is useful because it gives an insight into the interaction between an industry's market structure, in terms of firm sizes and numbers, and the innovation rate, and why and how these two may depend on the price elasticity of demand, barriers to entry, and the responsiveness of unit cost to R&D expenditure.

Suppose that uncertainty is absent and firms entertain Cournot conjectures, i.e. firm i, $i = 1, 2, \ldots, N$, chooses its level of output Q_i and R&D investment R_i as if these choices have no influence on the choices of the other $N - 1$ firms. Firm i makes its choice at the start of a (unit) period of time with a view to maximizing net profits in that period, i.e.

$$\text{maximizing}[\{p(\hat{Q} + Q_i) - c(R_i)\}Q_i - R_i] \tag{4.18}$$

where $\hat{Q}_i$ is the combined output of the other firms in the industry and $p(Q) = p(\hat{Q} + Q_i)$ is the industrial demand function. On the assumption that the solution to (4.18) is positive, it satisfies the first-order conditions, and these imply that

$$\frac{p - c}{c} = \frac{1}{\varepsilon N - 1} \tag{4.19}$$

$$\frac{R^*}{pQ^*/N} = \frac{\alpha}{p/c} \tag{4.20}$$

where Q^* represents the total industrial output, ε is the price elasticity of demand, i.e. $Q = \delta p^{-\varepsilon}$, and α, as in (4.14), is the elasticity of unit cost with respect to the R&D input. Equations (4.19) and (4.20) imply that

$$\frac{R^*}{pQ^*/N} = \alpha \left(1 - \frac{1}{\varepsilon N}\right) \tag{4.20*}$$

In the presence of barriers to entry N would be exogenously given, and the choice of R&D expenditure and output by the representative firm would then of course depend on N as well as on α and ε. We recall that the left-hand side of (4.19) represents the degree of monopoly and the left-hand side of (4.20) is the research intensity. It therefore follows that, as the number of firms in the industry falls, the degree of monopoly increases (equation (4.19)) but the research intensity declines (equation (4.20*)). If ε and α are assumed constant, (4.19) and (4.20*) can be solved explicitly in terms of Q^* and R^* (Dasgupta and Stiglitz 1980a: equations (4.32) and (4.33)). The solution implies that, with N falling, industry output also falls but the R&D expenditure by each firm increases. In the absence of any spill-over effect, it is that expenditure, and not the total R&D effort, that determines the rate of decline in unit cost, i.e. the industrial innovation rate. The model therefore implies that, with a decline in the number of firms in the industry and a consequent increase in the degree of monopoly, static efficiency also declines (because industry output falls) but innovation rate increases. This is precisely the Schumpeterian hypothesis.

There must be other models of inventive activity and market structure that would give rise to this hypothesis. In the Dasgupta–Stiglitz model the hypothesis is easily traced to the assumption that all firms either follow the same R&D strategy or, if following different strategies, do not share any of their inventive output. In such rather unlikely circumstances the rate of innovation in the industry is determined, not by the industry's total R&D effort, but by the R&D effort of one firm only. Accordingly, the R&D effort by the large firms in an industry with strong barriers to entry would be smaller as a proportion of sales, but larger in terms of its innovation effect, than the R&D effort by the smaller firms occupying the industry after the barriers have been reduced.

In the absence of barriers to entry the equilibrium number of (identical) firms N^* would be determined by the expectation that the maximum net profits, given by (4.18), would be driven by new entrants to zero, or near it. Hence

$$p - c = \frac{R^*}{Q^*/N^*} \qquad (4.21)$$

We have three unknowns, Q^*, N^*, and R^*, to be determined by three equations, (4.19)–(4.21), in which p is a function of Q^*, and c is a function of R^*. These three equations can be rearranged as follows:

$$\frac{p - c}{c} = \alpha \qquad (4.22)$$

$$\frac{R^*}{pQ^*/N^*} = \frac{\alpha}{1 + \alpha} \qquad (4.23)$$

$$N^* = \frac{1}{\varepsilon} \frac{1 + \alpha}{\alpha} \qquad (4.24)$$

Thus, in this model, the degree of monopoly and research intensity are decided solely by α, which is a technological parameter of the innovation production function. Relations (4.23) and (4.24) combined imply that the research intensity $R^*/(pQ^*/N^*)$ is proportional to $1/\varepsilon N^*$, i.e. in inter-industry comparisons involving varying industry-specific α and ε the research intensity would be inversely related to the price elasticity of demand but proportional to the degree of concentration. Variation in ε and α alone is suggested to be the cause underlying the two associations. When α declines, the cost mark-up and research intensity also decline, while at a given ε the number of firms increases. In the limit when $\alpha \to 0$, the Dasgupta–Stiglitz model reduces to the conventional model of a perfectly competitive industry. Again, because of the absence of any sharing out of the inventive output, industries in which α is small would tend to sustain large numbers of small firms and display low rates of innovation.

Moreover, (4.22)–(4.24) are just equilibrium relations. When the inflow of major product innovations is high, as it clearly is in most of the world today, giving birth to many new industries and rapidly shifting the demand in old industries, industries tend to be out of equilibrium all the time. Equilibrium conditions could then be expected to be at variance with the empirical evidence, the latter mirroring the constantly transitory non-equilibrium situations of industries. The only exception to this rule would be the industries where the speeds of adjustment to the rapidly changing environment are unusually fast.

(A technical point should be noted. It follows from (4.19) that, for the solution to exist, N must not be less than $1/\varepsilon$. However, for firm profits to be non-negative, N must not exceed N^*, given by (4.24). This defines the range of possible N in the presence of barriers to entry. The second-order maximization condition implies a constraint on ε and α, namely that $(1 + \alpha)/\alpha > \varepsilon$. Under a free-entry system yet another constraint on α and ε is implied by the requirement that no additional firm will find it profitable to enter the market. This condition is satisfied if either α or ε or both are chosen small enough.)

Process versus product innovation: the optimal mix under free entry

The model outlined above can be extended in a number of ways to the case when both product and process innovation can take place. One way is to suppose that the industrial demand still depends on the price alone, but each firm may influence the demand for its output, at a given price, through R&D investment in product innovation. The firm's net profit is then as follows:

$$\text{profit} = \{p(\hat{Q}_i + Q_i) - c(R_{1i})\}Q_i(R_{2i}) - R_{1i} - R_{2i} \quad (4.25)$$

The choice variables are the price p, the R&D expenditure R_{1i} intended to reduce unit costs, e.g. through process innovation, and the R&D expenditure R_{2i} intended to expand demand, possibly through product innovation. A change in R_{2i} would influence Q_i but have no effect on price. We shall assume that, in the long run, $Q_i \sim R_{2i}{}^\beta$. In other words, we presume that, given the price, the complete absence of product innovation by any given firm would cause its sales to gradually fall to zero. The firm also has the option of varying its output at a given R_{2i}, but then it must adjust the price to sell the output. Under free entry, the three first-order conditions and one zero-profit requirement give the following equations for symmetric equilibrium of the Cournot type:

$$\frac{R_2^*}{R_1^* + R_2^*} = \beta, \qquad \frac{R_1^* + R_2^*}{\text{sales}} = \frac{\alpha}{1 - \alpha - \beta} \quad (4.26)$$

$$\frac{p - c}{c} = \frac{\alpha}{1 - \beta} \quad \text{and} \quad N^* = \frac{1}{\varepsilon} \left(1 + \frac{1 - \beta}{\alpha} \right) \tag{4.27}$$

In these equations, β is the new parameter; it represents the elasticity of demand for firm output Q_i with respect to the firm's product-innovative R&D.

On comparing relations (4.26) and (4.27) with (4.22)–(4.24), it will be noticed that, given ε and α, both the cost mark-up and the research intensity are now higher, while the number of firms is lower. The research mix given by the ratio R^*_2 / R^*_1 is seen to depend solely on β, i.e. on the responsiveness of demand to the product-innovative effort (we have to assume of course that $\beta < 1$). The role of the parameters α and ε remains very much the same as in the original model. In particular, when α declines, given ε and β, the number of firms increases so that the industry becomes more competitive. At the same time, however, the firm's R&D expenditure and the industry's innovation rate both fall; the Schumpeterian negative relationship between static and dynamic efficiency thus remains in force.

Oligopolistic (with free entry) versus socially managed industry when the spill-over effect is present

In the presence of scale economies the socially managed industry would consist of a single firm. When the market demand is dependent on price alone (income effects absent) and the elasticities α and ε are constant, we can obtain an explicit solution for the optimum output and R&D expenditure (Dasgupta and Stiglitz 1980b: 273). This solution implies that the socially optimum research intensity equals α, which is to be compared with $\alpha / (1 + \alpha)$ under oligopoly with free entry. The socially optimum size of the R&D expenditure can be found to be less than the combined R&D expenditure of the competitive industry of the oligopoly type, but greater than the R&D expenditure of its representative firm. Consequently, under social management there is no waste due to the duplication of R&D effort (high static efficiency) and a greater innovation rate (higher dynamic efficiency).

To put it differently, in a competitive industry there is, from a social viewpoint, too much R&D effort but too little innovation. This important theoretical thesis is intuitively predictable, given the assumptions. Still, for anyone who has studied the innovation processes in centrally planned economies, this thesis must be a startling puzzle. There is therefore something fundamentally wrong with this part of the Arrow–Dasgupta–Stiglitz theory, a theme to which we return indirectly in Chapter 7.

However, some of the assumptions of this theory, perhaps the most crucial ones, cannot be easily relaxed. For example, it is widely thought that the pressures of more competitive environments tend to increase the efficiency of the R&D personnel and make firms' managements assimilate more of their own R&D output faster, as well as search more widely and with greater urgency for imitation opportunities. The models discussed in this chapter assume such things away. The difficulty is not of course in adopting a technology progress function which depends upon a suitable index of competitiveness, such as the degree of concentration. The real problem is in choosing a 'realistic' specification of any such relationship.

The assumption of no spill-overs made by Dasgupta and Stiglitz is easy to relax, however. One way of doing this is to suppose that a fraction λ of each firm's innovative output benefits each other firm. The cost function then has the form (4.14), where $\mu = 1 + (N - 1)\lambda$, and so it depends not only on one firm's R&D spending but also on how many firms there are. Our point will be easiest to make if we put $\beta = 0$ and $\varepsilon = 1$, and regard α as a constant. In this case the optimal solution under oligopoly with free entry has the form

$$N^* = \frac{1 + \alpha}{\alpha} \quad \text{and} \quad R^* = \sigma\alpha^2(1 + \alpha)^{-2} \tag{4.28}$$

where σ is a parameter of the market demand function: $p(Q) = \sigma Q^{-1/\varepsilon}$, so that, when $\varepsilon = 1$, σ represents the (constant) size of the market which our industry supplies. From (4.28) it follows that $R^* = \sigma/(N^*)^2$, and thus R^* is declining when N^* increases. The question is whether the presence of the spill-over effect would outweigh the effect of declining R&D expenditure by any one firm, and so result in the innovation rate increasing with N^*. On expressing R^* and α in terms of N^* and taking into account (4.14), we obtain the following relation between the unit cost and N^*:

$$\frac{\bar{c}}{c} = \sigma^{1/(N^* - 1)}(N^*)^{2/(N^* - 1)}\{1 + \lambda(N^* - 1)\} \tag{4.29}$$

The ratio $\bar{c}/c$ is a measure of the innovation rate. It will be noticed that, for large N^*, we have $\ln(\bar{c}/c) \approx (2/N^*)\ln N^*$ in the absence of the spill-over effect, and so $\bar{c}/c$ tends to zero as $N^* \to \infty$. However, when $\lambda > 0$ and N^* is large, then $\ln(\bar{c}/c) \approx 2\lambda \ln N^*$, and so it increases with N^*; introduction of the spill-over effect thus has the power to change the relationship between the rate of innovation and the degree of market competitiveness from negative to positive.

A note on strategic innovation

Modern developments of the models above consider the behaviour of competing firms in situations when a sequence of innovation decisions has to be made. A central question of this theory of strategic innovation is whether a technological leader can maintain its dominance, i.e. whether followers can leapfrog the leader or only gradually catch up. Given the early stage of development of the theory, we shall not include it in our discussion. However, Beath *et al.* (1990) give an excellent treatment.

Chapter five

Behavioural and evolutionary versus neoclassical theory of technical choice and innovation

Key principles of the neoclassical theory of technical choice

The contrast between the three rival theories indicated in the title of this chapter will be sharper if we recall briefly the key principles of the neoclassical theory. The latter may be the least realistic of the three, but is perhaps the best defined and the most developed. Its criticism also serves to define indirectly the other two theories.

The first and perhaps the most fundamental of the neoclassical principles is that all economic agents are presumed to be optimizers; they always can and do make choices which maximize their 'pay-off functions'. In particular, firms are supposed to select production techniques and output scales which maximize profits, while households demand consumption goods and supply labour hours in amounts which maximize their 'utility' levels.

Firms are also assumed to operate in an environment that is so intensely competitive that prices cannot and do not exceed unit total production costs for every good produced (goods in the case of joint output). However, whenever the prices actually charged result in loss of money by a production process (cost of inputs used exceeding the value of outputs produced), then this and every such process would instantly cease to be operated. This assumption is the second neoclassical principle, sometimes called the 'rule of profitability'. It ensures that neither losses nor pure profits (conventional profits net of the competitive return on capital) are ever made.

Prices are also assumed to be so highly flexible that the quantities demanded are always equal to the quantities supplied. This rule has one exception, however: in *quantity* terms, a good's supply may exceed demand for it only when its price has already reached zero level and therefore cannot fall further. This exception is the third neoclassical principle, called the 'rule of free goods'. It ensures that in each market, for gold as well as air, there would always exist a state of equilibrium between supply and demand in *value* terms.

As phrased above, the three principles represent an extreme and abstract version of the theory, referring as they do to the case of the perfectly competitive economy. In an economy that is less than perfectly competitive, individual markets may not be in equilibrium, but then changes in prices and quantities are presumed to take place in the direction of the market-clearing levels, and the speed with which such changes take place is, in a given market, related to the degree of competitiveness that is present in that market.

The choice of techniques under perfect competition in an n-sector economy

In an economy that is subject to perfect competition, prices, production techniques, and quantities of inputs and outputs would be determined jointly. To see the economic principles involved in this determination process, let us consider a simple example of an expanding closed economy in which n conventional goods (other than labour) are produced, each in its own multifirm industry. There may be a large number of alternative techniques available to produce each good, but for simplicity of exposition we assume that in the production of all goods only one (common) type of labour is needed. There are thus $n + 2$ prices to be determined: n good prices, the wage rate, and the rate of return on capital, which is defined as the ratio of the price of a capital good's services to the price of the capital good itself. The wage rate is common in all industries and the rate of return the same for all capital goods, by virtue of both competition and the absence of barriers to inter-industrial mobility of labour and capital resources.

According to the profitability rule, for any production technique in use the value of outputs must equal the value of inputs. Consider a set of n techniques, one from each industry's technology set. The cost–price equations, one for each technique as well as each good, give a system of n equations involving $n + 2$ prices. On eliminating the $n - 1$ good prices and taking the price of the nth good as unity, we obtain a relation between the real rate of return on capital and the real wage rate. Suppose that two different production techniques are available in each industry. Then 2^n different sets of techniques may be formed, each capable of producing all n goods and each giving rise to a relation between the real rate of return on capital and the real wage rate.

The envelope of these relations forms the so-called factor–price frontier. To each point of the frontier corresponds a set (occasionally a number of sets) of techniques which, given the real wage rate, maximizes the rate of return on capital, and another set or sets of zero pure-profit prices. (In the terminology of Chapter 1, these return-maximizing techniques are efficient and so belong to the production possibility frontier; thus the two frontiers are in a dual relationship.)

The real wage rate also influences the supply of labour services and, through the medium of a savings function, the rate of growth of the capital stock. The precise forms of these relations depend on the preferences of households with respect to goods versus leisure and on the distribution of income. The growth rate of capital stock may or may not be sufficiently high to absorb the labour services that are offered at a given real wage rate. If the economy is assumed to be in a state of full employment in the present period, a state of 'growth equilibrium' will exist only if the wage rate attains a value at which the growth rate of capital equals the growth rate of labour supplied in that period, so that the economy continues to enjoy full employment in the next period.

This growth equilibrium condition determines the wage rate, and thereby it closes the system of relations that determine, in the economy in question, the (return-maximizing and full-employment) set of techniques, one for each firm category.

The presence of technological progress changes little in the above analysis, provided that the progress is cost free and exogenously given. Simply, the appearance of new efficient combinations of techniques enables firms to obtain the same rate of return on capital at a higher real wage rate; thus the factor–price frontier shifts to the right. Technological innovation would also tend to shift the relation between the wage rate and the growth rate of the capital stock supplied, as well as the relation between the wage rate and the growth rate of the capital stock required to absorb the labour services offered at that wage rate (in Figure 5.1 these two relations are denoted $g_K{}^s$ and $g_K{}^d$ respectively).

Macroeconomic data on output and capital growth rates, as well as on wage rates and rates of return on capital, indicate that the innovation process tends to sustain g_K^* at a level that is both higher than the growth rate of employment and relatively stable over time, while producing a systematic upward rise in the real wage rate.

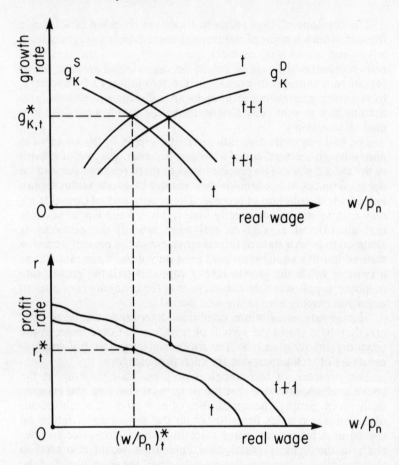

Figure 5.1 The relations determining competitive prices – the wage and interest rates included – and production techniques in periods t and $t + 1$.

Criticisms of the neoclassical theory

Most economists, including those who see themselves as belonging to the classical tradition, strongly emphasize the abstract unrealistic nature of the neoclassical principles enunciated above. Yet many see these principles as acceptable approximations of the actual tendencies: economic agents try to maximize their net economic benefits, prices tend to respond to excess demands or excess supplies, and the resultant price changes affect quantities supplied and demanded in the market equilibrium direction. At the same time it is widely accepted that firms do not attempt to maximize money profits only and 'that a goodly portion of all business behaviour may be nonrational, thoughtless, blindly repetitive, deliberately traditional, or motivated by extra-economic objectives' (Machlup 1951; cited by Machlup 1967; 5, footnote 1). It is also recognized, as in the Dasgupta–Stiglitz model discussed in Chapter 4, that the degree of competitiveness may differ substantially among industries, depending primarily on technological and demand characteristics and barriers to entry that jointly underlie particular market structures.

While a fuller assessment of the degree of error involved in applying the neoclassical principles is a subject in its own right, and very much beyond the scope of this chapter (see Kornai (1971) for one such assessment), it might be useful to identify the major difficulties of the theory, especially in the innovation context.

One such difficulty is that firms often have neither the information nor the computation capability needed to make profit-maximizing decisions, either operational (on current or short-term outputs, inputs, and prices) or strategic (on investment in human and fixed capital, and in R&D). The incomplete information and poor computation capability that firms do have 'precludes the use of any but simple rules of thumb as guiding principles' (Simon 1962: 10). Thus, even if firms wish to maximize profits, exactly or approximately, in many situations they do not because they cannot.

It was observed, for example, that if department stores were to allow profit maximization rules in their price-setting practices, they would have to pay close attention to (product-specific) marginal selling costs and price elasticities of demand. However, stores do not have such refined information. Consequently the rule that they follow instead is one of an approximately constant mark-up, in the United States typically between 40 and 45 per cent, for all the goods they sell (Cyert and March 1963; Baumol and Stewart 1971). Similarly simple non-classical rules are apparently also widely employed in the choice of major investment projects and in the determination of the firm's overall advertising budget (Simon 1962: 12; Baumol and Stewart 1971; 118, 139).

Apart from simple rules, recourse to 'informed guesses' and arbitrary judgements about future terms on which the business is to be conducted also seems as commonplace as it is inevitable. These 'future terms' are often random variables of unknown probability distributions; managers are not just uncertain about them, but simply ignorant. Examples are the future attitudes of the share-holders, the behaviour of competitors, the actions of government agencies, movements in wage and foreign exchange rates, and movements in commodity and other prices. The business decisions taken in such circumstances cannot be expected to be even 'optimally imperfect' (decisions in which the process of information gathering and calculation has been carried to the point of zero expected marginal net yield (Baumol and Stewart 1971: 119)), since usually there is no (sufficient) basis on which to form 'rational expectations' about the marginal net yield.

Natural selection and the evolutionary thesis

As informed guesses and judgements are bound to be management-specific, each industry is likely to be characterized by a considerable interfirm variation in the choice of firm strategies concerning the choice of production processes, range of products, scale of outputs, marketing techniques, R&D effort, management–union relations, and so forth, and by a parallel variation in profit performance. However, if firms operate in a competitive environment, a process of 'natural selection' takes place. Managements which by chance or as a result of extra qualities happen more often than not to make correct decisions, i.e. decisions which *ex post* turn out to be approximately profit maximizing, would tend to rise in standing and their firms would survive and possibly expand. At the opposite end of the performance scale the influence of unsuccessful managers would tend to fall, and their firms would be liquidated or taken over by the more profitable firms.

In the absence of innovation, an industry in which competitive pressures and forces of natural selection operate would thus be expected to evolve gradually along a path that converges towards the (neoclassical) optimal state. After a sufficiently long time, and this time would presumably depend above all on the degree of competition and the ruthlessness of the selection processes, the optimal state of affairs may well be a 'good' approximation of the actual state. The optimal solution is thus seen to be relevant for two reasons: first as a point which, in the space of variables that are subject to choice, 'attracts' through evolutionary process the actual choices at any time, and second as a reasonable approximation of these choices after a period of time.

The presence of innovation changes the perspective substantially without, however, making it completely irrelevant. The constant inflow of new ideas, changing fast the environment in which firms operate, makes each situation a new one, the past experience not always an asset, and the situation-specific optimal solution a moving target. Therefore the distance between the optimal state (again, the one which *ex post* turns out to be profit maximizing) and the actual situation may remain quite significant at all times, and so the former, under fast innovation, may never be a good approximation of the latter. However, as long as competition is present and selection processes operate, the moving target still retains its attraction power; in the space of quantities (of inputs and outputs) and prices it continues to signal the direction which the actual trend growth path of an industry is likely to follow.

The 'behavioural approach'

We have noted that the conventional (neoclassical) theory is useful in suggesting the way in which techniques of production, prices, and quantities may interact, and in explaining, in conjunction with the evolutionary thesis, the inter-industrial variation in major industrial characteristics, such as capital intensity and market structure. However, the theory is seen to be incapable of offering any interpretation of the usually wide interfirm allocative differences which we observe within each industry.

This intra-industry territory has been the subject of investigations by a number of industrial sociologists and organizational economists. In these investigations no *a priori* (maximizing or any other) behavioural rules are assumed. The ways in which the decisions are *actually* taken within the firm are studied instead. The results of these studies form the 'behavioural theory' of the firm.

The theory suggests that the firm should be viewed as a coalition of members, or sometimes groups of members, each motivated by different stimuli and attaching different weights to a potentially large number of goals which they all seek to attain. The goals themselves are to be found primarily in the areas of production, marketing, R&D activity, job satisfaction, and individual status and income, as well as in the firm profits, employment prospects, and growth. Each member also has somewhat different perceptions of the firm's internal and external environments in which the pursuit of the goals takes place, and of the constraint to which the members' strategies are subject.

Different firms may adopt similar organizational procedures for bargaining between the different interests to ensure a degree of consistency between the various goals, and for providing solutions to firm problems that would reflect the compromise. We cannot review these procedures or their rationalizations here. However, the point that is important for the purposes of this discussion is fairly clear. It is that both the actual decisions taken and the nature of their subsequent implementation must in part reflect the member-specific goals, knowledge, experience, and perceptions, and as these tend to differ considerably among firms, so do the firm decisions and profit performances.

Thus, while conventional theory makes reference to differences in the external environment in which firms of various industries operate, in explaining the inter-industry variation in *optimal* behaviour, the behavioural theory focuses on differences in the internal environments of particular firms in its explanation of the intra-industry variations in *actual* behaviour.

In the presentation above, however, the behavioural, evolutionary, and neoclassical theories of the firm are interpreted as essentially complementary rather than rival, with the evolutionary theory providing the link between the behavioural and the neoclassical theories.

The Nelson–Winter 'evolutionary' model of technical choice and innovation

The behavioural and evolutionary ideas are usually phrased in a manner that is too general to be suitable for conducting a (theoretical) analysis capable of generating specific propositions that could be subjected to empirical tests. For that purpose a further modelling effort is needed that would give the ideas an operational quality.

One such model has been proposed by Richard Nelson and Sidney Winter. They discuss its assumptions and some theoretical features, as well as reporting the results of its computer simulations, in a series of closely related papers published mainly in the 1970s and summarized and expanded in Nelson and Winter (1982b). The model is extremely simple, yet it has enough structure to offer some insight into the possible dynamics of the evolutionary process. Its presentation below is based on a variant that was used for simulation purposes (Nelson and Winter 1976). Some of the assumptions of this variant are very restrictive indeed, partly in order to preserve simplicity and focus on the non-standard ideas.

Suppose that an industry is made up of a large number of firms, all producing a common good and each employing the services of capital and labour, with both inputs being of the same type in all firms. There are also a large number of possible ways of producing the common good, each characterized by a pair of constant input–output coefficients a_L and a_K. However, in contrast with the neoclassical (standard) assumption, each firm initially knows only some of the production techniques, although all firms taken together may know all the techniques.

In each time period the state of a given firm j is fully described by the firm's technique $(a_L, a_K)_j$ and by its scale of operation, which will be indicated by its capital stock K_j. This capital stock is assumed to be always fully utilized. Thus the demand constraint is presumed not to operate, even though the good price, expressed in terms of the aggregate price of all other goods, remains constant. Moreover, if a firm happens to make losses at full-capacity output, one of its responses will be to disinvest by selling some of its capital assets (see below) while still producing at full capacity. The full-utilization assumption implies that the aggregate (industrial) capital stock, output, and employment are respectively

$$K = \Sigma \ K_j \qquad Y = \Sigma \ \frac{K_j}{a_{K_j}} \qquad L = \Sigma \ \frac{a_{L_j}}{a_{K_j}} \ K_j \qquad (5.1)$$

The wage rate w is the single endogenous price in the model. It is set at a level just high enough to ensure that the industry attracts enough labour to meet its requirements. Hence

$$L^s(w) = L \qquad (5.2)$$

where $L^s(w)$ is the labour supply function.

In a transition from any one (unit) time period to the next, the capital stock of a firm with a positive capital in the current period is increased by net investment, assumed to be equal to the excess of net profits (profits minus capital depreciation) over and above the 'required dividend' which is taken to be a given percentage of the capital stock; all excess net profits are thus always reinvested. This assumption can be written as follows:

$$K_{j,t+1} - K_{j,t} = pY_{jt} - w_t L_{jt} - \delta K_{jt} - DK_{jt}$$

$$= \left(\frac{p - wa_{L_j}}{a_{K_j}} - \delta - D \right)_t K_{jt} \qquad (5.3)$$

where δ stands for the depreciation rate and D represents the dividend rate. It will be noticed that the size of the net investment by each firm in period t is dependent not only on the production technique and size of the particular firm in that period, but also on the production techniques and sizes of all the other firms, for they and the firm in question determine the common wage rate jointly, and thereby all the firms together influence the costs and profits of each individual firm. The endogenously determined wage rate and

73

the exogenously given dividend rate thus represent a price system in the model that plays its usual role of a mechanism for re-allocating resources away from firms with low profit rates to those with high rates.

Apart from the essentially random choice of the initial production technique, the spirit of the behavioural approach is captured mainly by the assumptions about the innovation process, i.e. about the way that firms attempt to identify and adopt better techniques. In this area of behaviour the firms are presumed to be guided by four simple rules.

(i) First of all, firms cannot and do not maximize profits in the neoclassical fashion. Following Herbert Simon (1957), they are assumed to be *satisficers* instead.

Accordingly, the presently employed production technique would continue to be used as long as it earns a 'satisfactory' profit rate and irrespective of the potential profit rates that alternative production techniques, from among those identified, would have promised. Only a drop in the actual profit rate below a 'minimum satisfactory' level triggers off a serious search for an alternative technique, with the expressed aim of replacing the current technique. The firms undertake no R&D activity except the occasional search which, in this particular model, is assumed to be costless.

(ii) The probability $p_s(h')$ per unit time of identifying a technique h' by a firm using technique h during a 'local search' is assumed to be a declining function (linear in the simulations reported) of the 'distance' between h and h', i.e.

$$p_s(h') = \alpha - \beta D(h, h')$$

where the distance

$$D(h, h') = \pi \left| \log \left(\frac{a_K{}^h}{a_K{}^{h'}} \right) \right| + (1 - \pi) \left| \log \left(\frac{a_L{}^h}{a_L{}^{h'}} \right) \right| \qquad (5.4)$$

This assumption is one possible specification of the notion that minor in-house improvements are easier to discover than major departures from the technique currently used. There are two parameters of the specification, π and β. A value of π above 0.5 indicates that it is inherently more difficult to identify a given percentage reduction in the capital–output coefficient than in the labour-to-output ratio, and vice versa when π is below 0.5. It will be noticed that this bias parameter is not a subject of choice for the firms in question but a characteristic of their innovational environment. The 'easiness' parameter β in Figure 5.2 is another such

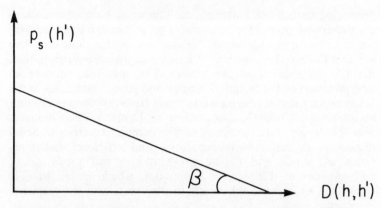

Figure 5.2 The probability of finding technique h' through a local search is negatively related to its technological 'distance' from the technique h currently in use.

characteristic; it indicates the easiness in identifying major in-house improvements with input–output coefficients significantly different from the current ones.

(iii) A searching firm may also note the existence of another technique, with a view to imitating it, if the technique is already employed by some of the other firms. The probability $p_{im}(h')$ that it will in fact identify a particular technique h' is assumed to be positively related (proportional in the simulation exercise) to the share of output accounted for by the firms using that technique in the current period.

The combined probability of discovering technique h' is then

$$p(h') = \lambda p_s(h') + (1 - \lambda)p_{im}(h') \tag{5.5}$$

where a high value of λ represents a firm which prefers to engage in local search rather than in imitation.

(iv) In the next period the firm will adopt with probability $p(h')$ the technique h' only if it meets a profitability test by promising to yield a higher rate of profit than that yielded by the technique currently employed. The model also allows for the firms in business to misjudge the input coefficients of the alternative techniques and for new entrants to appear.

In this particular version, the model has two behavioural parameters, D and λ, and two environmental parameters, π and β. Nelson and Winter select two values for each parameter, thus producing sixteen combinations, each implying both an evolutionary pattern of growth for firms and a growth path for the industry (or the economy) as a whole. Nelson and Winter show that they are able to reproduce, with a rather high degree of approximation, the US historical aggregate growth data, and thus provide an interpretation of the technical change and growth processes in the US economy that is alternative to that offered by the neoclassical approach *à la* Cobb–Douglas, Solow, or Denison. Their model is also able to generate the firm-size distribution, the cross-sectional dispersion in capital-to-labour ratios and efficiency indicators (such as labour and capital productivities and profit rates), and patterns of diffusion of techniques, which are all broadly consistent, or at least not clearly inconsistent, with the empirical evidence.

Two behavioural features of the model are crucial. One is inspired by the notion that the knowledge about alternative techniques is firm-specific and highly incomplete. This idea is to some extent captured in the model by the device of a probabilistic transition from one technique to another, subject to meeting the profitability test. The effect of it is that the same firm, or two seemingly identical firms, may adopt widely different techniques in the same circumstances.

The other important feature is that firms do not innovate when their profitability performance is above the minimum satisfactory level, and search for alternative techniques only under the threat of immediate contraction. This clearly myopic behaviour may well be realistic in some cases but is probably unnecessarily restrictive as a general proposition. Highly profitable firms presumably wish to maintain their favourable position, and also have the resources to do so. The industrial experience helps them to foresee the potentially dangerous implications of innovational inaction. It is therefore not surprising that the empirical evidence which we reported in Chapters 2 and 3 indicates the presence of a positive relation between firms' profitability and the intensity of their R&D activity. This evidence runs very much against the satisficing principle, thus to a considerable extent undermining the basis of the Nelson–Winter theory.

We have also noted that one of the key characteristics of the innovation and growth processes appears to be the presence of a strong variation among industries in the innovation potential. The empirical vindication of the Mansfield model (Chapter 2) suggests that firms allocate their R&D resources above all to the pockets of economic activity where they expect to obtain particularly high returns on their R&D investment. This behaviour is a further reason why firms in certain sectors, such as aviation or electronics in the post-war period, have been at the same time highly innovative, very profitable and fast expanding. The Nelson–Winter firms innovate only when they expand slowly or, indeed, are contracting.

Fortunately, the model has the standard evolutionary feature that firms which (during the periods of low profits and slow growth) happen to be unsuccessful in innovation would be reduced in size and possibly eliminated. The selection mechanism thus operates, pulling the firms that remain in business in the direction of the production possibility frontier. At the same time the Nelson–Winter firms are identical in terms of the modes of behaviour of their managements with respect to the choice of techniques and innovation strategies. The interfirm variation in that behaviour is purely random. The behavioural characteristics are not firm-specific, but change *ad hoc* from period to period. The essential feature of real economies, namely that managers vary in their entrepreneurial abilities, is thus lost. For this reason the Nelson–Winter theory cannot be said to be really Schumpeterian. This variation among firms is an in-built insurance mechanism against failure of the industry as a whole. The simultaneous testing of several alternative strategies over a period of time increases the chance of discovering a successful strategy. Subsequently the successful managers increase in power and the successful firms increase in size. Variation and selection are thus two sides of the same coin of a truly evolutionary process. A theory which does not allow for such inherent variation cannot be said to be really evolutionary.

It will be noticed that the production possibility frontier is implicitly present in the model. The production techniques which the authors picked up randomly, and which are out there for the firms to find, are supposed to be God-given. This set of techniques is constant over time, and those among them which are Pareto efficient represent the ultimate (potential) production possibility frontier. At any given time, however, only some of the techniques of the set would have been identified by one or more firms. These techniques form a subset of the full set. The subset is enlarging over time, as new techniques are being identified. There are always likely to be some firms in the model which are well inside the subset, but with time these firms either disappear or move towards the production possibility frontier. Firms are of course profit improvers rather than maximizers, but the industry does tend to converge, although slowly, towards the neoclassical optimum, which itself is moving.

The Nelson–Winter model, and indeed any evolutionary or other non-neoclassical theory, is important in so far as the process of reaching the optimum is of interest rather than the optimum itself.

Chapter six

Innovation diffusion: theory and evidence

The term 'diffusion' is widely used in many fields of science, where it refers essentially to the same phenomenon: the spread in space or acceptance in a human environment, over time, of some specific item or pattern. The phenomenon in question may be the spread of matter of a given type in physics, of an idea or pattern of culture in anthropology, of a practice or an institution in sociology, or of a product or a method of making one in economics. The primary interest of the (theoretical and empirical) diffusion studies in all these fields is equally common: to identify the factors that influence (facilitate or resist) the process of spread and then to discover the precise relation between these factors and the rate of spread within a given environment. There are also further, although more distant, similarities, some of which will become apparent in the course of this chapter. However, unlike physical phenomena, which do not need man to take place, anthropological or economic diffusion processes involve people in an essential way, as both diffusion agents and adopting units. This fact complicates the task of a social scientist enormously, since human abilities, values, and attitudes immediately come into play, and these are inevitably highly heterogeneous and essentially situation-specific.

In the previous chapter we emphasized that economies are typically out of equilibrium. However, if initially any economy happened to be in equilibrium, then an act of innovation by a firm, its component, would have represented a disturbance, since presumably the firm would as a result perform better than before, and possibly better than the other firms. This difference in performance may in turn represent an incentive for some or all the other firms to adopt the innovation as well. The initial innovation act may thus be the start of an innovation process. The process is diffusional for any particular innovation and transitional for any particular economic system, taking the latter from an initial to a new equilibrium state.

79

The economic system in question is usually an industry in a ...en country, but it may also be a multi-unit firm, a country, or the world as a whole. Depending on the system involved, we speak of a diffusion process that is intrafirm, intra-industry, inter-industry, or international. The macro-oriented analysis of innovation and economic growth in the presence of international diffusion, or international technology transfer, is the primary subject of Part II of this book. The microeconomics of diffusion is now a fairly large subject, and good recent treatments are given by Stoneman (1983) and Thirtle and Ruttan (1987). In this chapter we shall illustrate the various strands of it by discussing three important and substantially different models: those of Edwin Mansfield (1961), Stephen Davies (1979), and Peter Grindley (1986). We also discuss in some detail the so-called epidemic model, a forerunner of nearly all diffusion theory. The particular aspect we shall look at in these three models is, among other things, the distribution of roles between the economic and non-economic factors in influencing the rate of diffusion. We shall also make note of some of the major empirical findings concerning the significance of these factors.

Two key stylized facts of innovation diffusion

The usual measure of the extent to which an innovation has been diffused within an industry, in the case of a new process inno-vation, is the flow of output based on the new process, given as a percentage of the maximum flow of output in the industry that potentially can be produced using the process. In the case of a new product, the extent of diffusion can be measured by the volume of the new product in use, given as a percentage of the maximum stock of the new product that potentially can be used. The latter measure is also referred to as the rate of displacement. This rate is to be distinguished from the rate of imitation, which is usually understood to be the proportion of all potential users in a particu-lar industry or economy that have adopted the innovation. Thus the rate of imitation ignores the extent to which the innovation is actually used by the units that have adopted it.

Two important features of the diffusion phenomenon were noted by early investigators, such as Ryan and Gross (1943), and have since continued to command a great deal of attention. One is the very wide range of rates at which different innovations spread and the other is the general S-shaped or logistic-type pattern of diffusion that most innovations apparently have in common. These are now often regarded as key stylized facts of the diffusion process. Therefore most of the research in this field, both theoretical and empirical, tends to centre on identifying those characteristics of innovations, and of the economic and human environment in which their diffusion takes place, that contribute most to the variation in diffusion rates and the uniformity of diffusion patterns.

Mathematical theory of spread of information and the logistic curve

One possible theoretical approach to diffusion rests upon the view that any innovation may be regarded as a piece of information which spreads by contact through a population of potential users. The mathematical theory of contagion and spread of information can then be applied to give the rate and extent of spread at any given time. Most of this theory was developed before serious study of innovational diffusion started, and it has been useful for economists in listing clearly the assumptions under which the accumulated spread follows the logistic pattern. It is worthwhile reproducing these assumptions, if only to realize how stringent they are and hence that, in the real world, deviations from the purely logistic pattern are only to be expected. They are as follows:

(i)   the number n of individuals (firms) in the total population is constant during the spread of information, the nature of which does not change;

(ii) the frequency α of contacts in unit time between any two individuals is constant over time and the same for all pairs;

(iii) the probability p that the information (disease, innovation) will be transmitted when a contact occurs between an affected and an unaffected individual is constant over time (in particular, independent of the age of the information and of the time period that the affected individual had it) and is the same for all such pairs;

(iv) once acquired, the 'state' (disease, innovation) cannot be lost.

If $x(t)$ is the number of affected individuals at time t, then by (ii) $\alpha x(n - x)\Delta t$ will be the number of contacts between the affected and the unaffected individuals in the period from t to $t + \Delta t$. Bearing in mind assumptions (iii) and (iv), it follows that

$$\Delta x(t) = \alpha p x(t)\{n - x(t)\}\Delta t \tag{6.1}$$

Let $z(t)$ denote the fraction $x(t)/n$ of the population affected. In terms of z, equation (6.1) has the form (we replace $\Delta x/\Delta t$ by dx/dt, or $\dot{x}$)

$$\dot{z} = \beta z(1 - z) \qquad \beta = \alpha p n \tag{6.2}$$

81

which has as its solution the logistic function

$$z(t) = \frac{1}{1 + \exp(-a - \beta t)} \qquad (6.3)$$

where a is a constant of integration defined by the given $z(0)$. From (6.3) it follows that

$$\ln\left\{\frac{z(t)}{1 - z(t)}\right\} = a + \beta t \qquad (6.4)$$

which is the form generally used for the purpose of estimating the coefficients a and β, given the observations on $z(t)$.

From (6.4) we can immediately find the time interval needed for the fraction z to increase from a level z_0 to z_1:

$$\Delta t = \frac{1}{\beta} \ln\left\{\frac{(1 - z_0)/z_0}{(1 - z_1)/z_1}\right\} \qquad (6.5)$$

Thus the interval is inversely related to the parameter β, which is the reason why β is widely used as a measure of the speed of the diffusional process. Since $\beta = \alpha p n$, the model identifies the frequency of contacts α, the probability p that any such contact will be effective, and the number n of potential adopters as the factors which determine the speed of the diffusional process. The probability p may be influenced by the more basic parameters that characterize both the innovation in question and the adopters. However, given p, the rate β is predicted to be proportional to $\alpha(n)n$. Since α is likely to be related positively to n, the model predicts, under a *ceteris paribus* clause, that the spread of innovations would be (much) faster in less concentrated industries (more firms of a similar size supplying the same total output). Since the model emphasizes the number of effective contacts as the speed-determining factor, this prediction is not surprising; to use an illustration, epidemics should spread faster in London or New York than in the countryside.

However, assumptions (i)–(iv) listed above cannot plausibly be claimed to represent a 'reasonable approximation' to industrial reality. In the context of innovation diffusion, especially in modern industry, contact alone may well be relatively unimportant. Advertisements and professional publications may alert and inform all the potential adopters of an innovation at almost the same time, and yet the firms may vary so much in their individual circumstances (the age of capital, the product unit, the size, the human skills, the financial position) that their response to the innovation may still be far from uniform. In terms of the model above, the probability p may vary considerably among firms. Moreover, in many actual situations the volume (number) of potential applications (users) of an innovation may change rapidly; the innovation itself may undergo major improvements during the diffusional process, thereby making it more attractive; the rapidly changing environment of prices and demands may greatly affect the profitability of the innovation in question and hence also the probability that it will be accepted; the spread of the innovation may also be halted by the invention of an attractive substitute (monochrome and colour television would be an example).

However, assumptions (i)–(iv), while sufficient, are not all necessary to generate a logistic-type process. In equation (6.2) we only require β and n to be constant. Since $\beta = \alpha pn$, this requirement is consistent with changes over time in α and p. More importantly, there may be radically different models (sets of assumptions) capable of producing diffusion patterns that would also be S-shaped or even of the logistic type. For example, in the initial state of the diffusion process the rate of spread, to be measured in terms of Δx or Δz per unit of time, may be low simply because of the prototype nature of the innovation rather than because of the small number of 'affected' firms. This rate could be increasing in the middle stage of the process as subsequent improvements of the innovation make it more attractive, but slow down in the mature stage when the potential for most useful applications of the innovation would have been largely exhausted. The S-shaped pattern may thus mirror, at least in part, a particular sequence of improvements and applications – the sequence described in Chapter 1 under the product cycle hypothesis. (Improved innovation, strictly speaking, is a new innovation, but presumably it is an innovation of the same type. Therefore most empirical studies of diffusion are concerned with the same innovational types rather than individual innovations.) Another such example is the class of consumer goods, such as durables, the demand for which is strongly dependent on income. In this case, the S-shaped pattern of spread would be generated by a steady shift with time of the bell-shaped distribution of real incomes.

The Mansfield model

Another possible way of explaining the S-shaped pattern of diffusion was suggested by Edwin Mansfield (1961). His model makes no reference to the product cycle theory or demand functions, but is rather closely related to the theory of epidemics, the simplest (stationary) case of which was discussed above. To see this closeness more clearly, let us rewrite equation (6.1) as follows:

$$\frac{\Delta x}{n - x} = \alpha p x \, \Delta t = \beta \, \frac{x}{n} \, \Delta t \qquad (6.6)$$

That is, under assumptions (i)–(iv) the number Δx of new adopters in the time interval from t to $t + \Delta t$, taken as a proportion of all 'hold-outs' $n - x$ at time t, would be equal to the number $\alpha x \Delta t$ of potentially contaminating 'contacts' that each firm makes multiplied by the probability p that the innovation will actually be transmitted when a contact occurs. Mansfield begins the development of his model by arguing that the left-hand side of (6.6) may be any increasing function of x, but eventually the function is specified as linear. Mansfield thus regards (6.6) as an acceptable approximation of the true equation for the spread of innovation. However, he redefines the term 'contact' and specifies the probability p as a function of certain economic variables. The term contact is in fact never used. Instead, his assumption that the ratio $\Delta x/(n - x)$ should be positively related to the proportion of the firms already using an innovation is made on the following grounds:

> As more information and experience accumulate, it becomes less of a risk to begin using it. Competitive pressures mount and 'bandwagon' effects occur. Where the profitability of using the innovation is very difficult to estimate, the mere fact that a large proportion of its competitors have introduced it may prompt a firm to consider it more favourably. Both interviews with executives in the four industries and the data indicate that this is the case.   (Mansfield 1968: 137–8)

There is nothing in this argument that would necessarily imply proportionality between $\Delta x/(n - x)$ and x/n, something which in Mansfield's model is essentially an *ad hoc* assumption adopted for reasons of mathematical convenience. The proportionality could be justified, but then the stringent assumptions (i)–(iv), or perhaps another set of clearly stated assumptions, would have to be adopted explicitly.

The variables which are considered to have an influence on the probability p include the expected profitability π of introducing an innovation, the size S of the investment required to install it, and the annual rate of growth g of industry sales. However, the second crucial assumption of the model is that these innovation-specific variables remain constant during the diffusional process. The role of innovational improvements, which is rather critical in the case of many innovations, is thus downplayed.

The solution for $z_i(t)$, where i numbers the innovations which are spreading in a particular industry, could thus still be given by equation (6.4). Alternatively, if we have observations on $z_i(t)$, we can estimate the parameters a_i and β_i. Having obtained the estimates of β_i, i.e. the innovation-specific rates of diffusion, we can proceed with the second stage of analysis in which the variation in β_i would be explained in terms of the variations in π_i, S_i, g_i, and possibly other variables. In Mansfield's studies a linear form is typically assumed:

$$\hat{\beta}_i = a_{0i} + a_{1i}\pi_i + a_{2i}S_i + a_{3i}g_i + \ldots$$

where $\hat{\beta}_i$ stands for the estimate of β_i. The usual statistical tests are then employed to judge the explanatory power of the model. For example, Mansfield (1961) finds this power to be rather impressive for the admittedly small sample of twelve innovations in the four US industries he studied. Later in this chapter we shall contrast this model with some of the other models of diffusion, emphasizing in particular differences in assumptions concerning the behaviour of the firm(s) concerned.

The Davies model

The empirical literature has long called attention to the fact that in many, perhaps most, cases of innovational diffusion, the innovation under study undergoes substantial improvements in the course of the diffusion process. All the while, as a result of such improvements, its value to potential adopters grows. At the same time this value could differ substantially among firms. We have already argued informally that when these two features are present the diffusion curve will be S-shaped. We have also noted that Mansfield's model is rather inflexible in that it assumes the same profitability of an innovation for all firms not only at any given moment of time but also through time. The primary purpose of Davies's model is to relax these two assumptions of the Mansfield model.

The essential features of his theory (Davies 1979) are as follows. A capital goods industry offers to supply an innovation, embodied in capital equipment, to a user industry comprising n potential adopters. Patent restrictions do not operate and each firm knows something about the innovation from the start. However, the quality of the information about the technical and economic aspects of the innovation may initially be poor, and improve only with accumulation of experience within the industry and when the competitive pressures originating from adopters induce non-adopters to engage in active search. Moreover, the understanding of the innovation and the grounds for its adoption (rejection) may vary considerably between firms.

A particular firm i, of size S_{it} at time t, is assumed to be using the innovation only if the expected pay-off period R_{it} is less than or equal to the maximum pay-off period R_{it}^* acceptable to i. These two pay-off periods are supposed to be firm-specific. They are specified as follows:

$$R_{it} = A_t^e S_{it}^{\beta_1} \varepsilon_{1it} \qquad R_{it}^* = A_t^* S_{it}^{\beta_2} \varepsilon_{2it} \qquad (6.7)$$

A_t^e and A_t^* reflect those changes in R and R^* over time that are common to all firms, while ε_{1i} and ε_{2i} reflect the particularities of each individual firm. A^e is assumed to fall with t, mainly because of subsequent improvements in the innovation. A^* is also expected to fall, because the improvements in the quality of information about the innovation reduce the risk premium component of the target profitability rates. Competitive pressures that originate from the adopters may have similar effects on R^*.

From (6.7) it follows that

$$\frac{R_{it}^*}{R_{it}} = A_t S_{it}^{\beta} \varepsilon_{it} = \left(\frac{S_{it}}{S_{it}^*}\right)^{\beta} \qquad (6.8)$$

where $\beta = \beta_2 - \beta_1$, $A_t = A_t^*/A_t^e$, $\varepsilon_{it} = \varepsilon_{2it}/\varepsilon_{1it}$ and $S_{it}^* = (A_t \varepsilon_{it})^{-1/\beta}$. Assume that $\beta > 0$ (the model with $\beta < 0$ is symmetrical). The innovation is adopted when $R_{it}^*/R_{it} \geqslant 1$, or when $S_{it} \geqslant S_{it}^*$; S_{it}^* can be interpreted as the threshold or critical firm size. Since S_{it}^* is a random variable, it is the probability that $S_{it} \geqslant S_{it}^*$, in short $P(S_{it} \geqslant S_{it}^*)$, that defines the probability that firm i will use the innovation at time t. If ε_{it}, at each time t, is assumed to be a multiplicative function of the various firm-specific random characteristics, all independent of each other and normally distributed, the central limit theorem implies that $\log \varepsilon_{it}$ will be normally distributed across firms in every period. The mean value of ε_{it} and the variance of $\log \varepsilon_i$ are assumed to be common to all firms, i.e. to be industry-specific rather than firm-specific; the former is normalized to unity and the latter is denoted by σ_t^2. Therefore it follows from the definition of S_{it}^* that $\log S_{it}^*$ is also normally distributed and that the mean value and the variance of $\log S_{it}^*$ are $-(1/\beta) \log A_t$ and $(\sigma_t/\beta)^2$ respectively. Hence the probability that the critical size S_{it}^* is actually not more than the given size S_{it} is a cumulative log-normal function of S_{it}, with the same mean value $-(1/\beta) \log A_t$ and variance $(\sigma_t/\beta)^2$.

The firms in our user industry must also be distributed with respect to size in some way. Davies adopts the standard view that the firm-size distribution is also typically log-normal. The density distribution $\mathrm{d}F_t(S)$, which stands for the frequency of cases when the actual size S_{it} falls into a given range from S_t to $S_t + \mathrm{d}S_t$, is then $\mathrm{d}\Lambda(S_t; \mu_{st}, \sigma_{st}^2)$, where μ_{st} is the mean and σ_{st}^2 is the variance of the size distribution.

The product $P(S_{it}^* \leqslant S)\mathrm{d}F_t(S)$ thus gives the probability that all firms of sizes in the range from S to $S + \mathrm{d}S$ would use the innovation. Hence, by summing over all firm sizes we obtain the probability that a randomly selected firm has adopted the innovation. This probability, denoted below by Q_t, is a stochastic equivalent of the proportion x_t/n of firms which have adopted the innovation, and is therefore used by Davies as a measure of the extent of diffusion within the industry at time t. Thus

$$
\begin{aligned}
Q_t &= \int_0^\infty P(S_{it}^* \leqslant S) \, \frac{\mathrm{d}F_t(S)}{\mathrm{d}S} \, \mathrm{d}S \\
&= \int_0^\infty \Lambda\left\{ S_t; -\frac{1}{\beta} \log A_t, \left(\frac{\sigma_t}{\beta}\right)^2 \right\} \mathrm{d}\Lambda(S_t; \mu_{st}, \sigma_{st}^2) \\
&= N\left\{ \frac{(1/\beta) \log A_t + \mu_{st}}{(\sigma_{st}^2 + \sigma_t^2/\beta^2)^{1/2}}; 0, 1 \right\} \\
&= N\left(\frac{\log A_t - \mu_D}{\sigma_D}; 0, 1 \right) \\
&= N(\log A_t; \mu_D, \sigma_D^2) \quad\quad\quad\quad\quad (6.9)
\end{aligned}
$$

where use is made of the 'convolution theorem', from which

$$
\int_0^\infty \Lambda(y; \mu, \sigma^2) \, \mathrm{d}\Lambda(y; \overline{\mu}, \overline{\sigma}^2) = \Lambda(1; \mu - \overline{\mu}, \sigma^2 + \overline{\sigma}^2)
$$

and the definitional identity

$$
\Lambda(y; \mu, \sigma^2) = N(\log y; \mu, \sigma^2) = N\left\{ \frac{\log(y - \mu)}{\sigma}; 0, 1 \right\}
$$

Moreover, in equation (6.9),

$$
\mu_D = -\beta\mu_{st} \quad\quad \sigma_D = \sigma_t^2 + \beta^2\sigma_{st}^2 \quad\quad\quad (6.10)
$$

In the case when the distribution parameters remain constant over time, the function Q_t is seen to follow a cumulative normal time path only when $\log A_t$ is proportional to time, which is the case when A_t is increasing exponentially. This is a strong requirement, and therefore the model predicts the pattern of diffusion to be of the logistic type only as a very special case. However, the incidence of a general S-type diffusion pattern would still remain high.

Compared with Mansfield's model, three new parameters have come into play: μ_s, σ_s, and σ. Given t and A_t, equation (6.9) implies that the fraction Q_t is positively related to $(\log A_t - \mu_D)\sigma_D$. However, the latter is higher, the lower μ_D and σ_D are. Therefore, in an inter-industry comparison involving industries with the same A_t, the cumulative diffusion would be higher in industries in which the (size and other) differences among firms are smaller and, given σ_D, where the mean size of the firms is larger. If μ_D and σ_D are given, the crucial role is played by a number of dynamic factors, common to all firms in a given industry, which underlie function A_t in equations (6.8) and (6.9).

It is interesting and somewhat paradoxical that when A_t is constant over time, perhaps after rising initially, the fraction Q_t would have stopped increasing at a level below unity, even though the value of A_t may be high (assume $\sigma_D \geqslant 0$ and both μ_D and σ_D constant). This property of equation (6.9) is a consequence of the model's allowance for the possibility that firms discard the innovation whenever the special technical or financial circumstances, captured by ε_{it} in equation (6.8), reduce firm i's expected profitability below its minimum (target) level, which for $\sigma_D > 0$ will occur with positive probability.

It should also be noted that when firms are identical in every respect, so that $\sigma_D = 0$, the industry would switch instantly from no firms adopting to all firms adopting as soon as the sign of $\log A_t - \mu_D$ changes from negative to positive. This off–on pattern of diffusion is of course the one which the 'pure' neoclassical theory of the firm would predict. The fact that actual diffusional processes usually proceed slowly is therefore indicative of the typically high degree of heterogeneity of firms, of which size is just one dimension.

Davies's model can also be used to supply an interpretation of the observed tendency for the average speed of intra-industry diffusion to have increased considerably over the last two centuries or so. According to this model, this decline may in part be due to changes over time in the dynamic factors underlying A_t: in modern times the post-invention improvements are possibly made faster; innovation producers provide and potential adopters may search for information more extensively and more effectively; competitive pressures on non-adopters may be greater. However, equation (6.9) suggests that there might also have been another, and perhaps less obvious, factor at work: the spread of professionalism among managers and the development of a sophisticated infrastructure of financial institutions, ready to offer finance and advice. This factor could have reduced the variation among firms in the target rates of profit required by them and in the methods of evaluating the expected profitability of any investment project. As a result, the variation in R_i^*/R_i among firms may have declined, thus lowering the variance σ_D^2 in equation (6.9) and hence increasing the likelihood that in each period there will be a greater number of firms responding in the same way to an innovation (by adopting or non-adopting).

Some empirical findings

The detailed studies of diffusion, such as those reported by Nasbeth and Ray (1974), tend to emphasize the large number of firm- and innovation-specific factors that appear important in explaining the timing and extent of adopting an innovation. Many of these factors, though potentially important, such as those underlying the 'innovative dynamism' of the managers or resistance of the labour force, are difficult to identify and almost impossible to quantify. Most econometric studies have followed Mansfield in first estimating a measure of the rate of diffusion of a particular innovation in a particular industry and then regressing it or its logarithm against a number of likely explanatory variables, usually across innovations for a given industry or across industries for a given innovation. In all these studies the average rate of return expected from adopting an innovation almost invariably emerges as a major, possibly the key, variable explaining the rate of diffusion. However, firms are apparently often highly uncertain about the probable opportunity costs and benefits associated with investment in new innovations. According to an earlier UK study, as many as 64 firms in a sample of 116 used 'commercial acumen', hunches, or crude rules of thumb for evaluating the profitability and commercial viability of their innovation projects (Carter and Williams 1957: Chapter 5). The size of firms tends to have a positive influence, as in the case of all twenty-two process innovations studied by Davies (1979: 166). However, size may also be a proxy for some other variables, such as the quality of management and R&D capability, and so capture their combined influence as well. Economic variables indicating low risk (a low cost of adopting as a proportion of firm's assets) and high competitiveness (a large number of firms similar in terms of size and other characteristics) are also found to stimulate diffusion (Mansfield 1968; Romeo 1975).

Davies's more sophisticated model, while useful for the purpose

89

of broadly interpreting the observed variety in the overall patterns of diffusion – when and why they differ from the logistic curve – uses a number of unobservable variables and for this reason represents a less obvious basis for empirical work. The function A_t in equation (6.9) is clearly a 'black box'; the relation between A_t and the underlying 'dynamic factors' is not specified and therefore the size of the separate contributions of these factors to the rate of diffusion cannot be identified. Yet it is largely A_t which in this particular model decides the overall pattern of diffusion. At the present stage of sophistication, the main use of the model is for finding the best (simple) specification of the function A_t for a particular innovation or a group of innovations. Davies himself divides his sample of twenty-two innovations into two such groups. For one of these groups he supposes that $A_t = \alpha \exp(\psi t)$. Also, let σ_s and σ be constant, but let the size of all firms change with time at a proportional rate δ, so that μ_{st}, the logarithm of the mean size, changes in a linear manner: $\mu_{st} = \mu_{s0} + \delta t$. On substituting these two specifications we have

$$\frac{\log A_t - \mu_{Dt}}{\sigma_D} = \frac{t - \mu_D^*}{\sigma_D^*} \equiv y_t \tag{6.11}$$

where $\mu_D^* = -(\log a + \beta\mu_{s0})/(\psi + \beta\delta)$ and $\sigma_D^* = (\sigma^2 + \beta^2\sigma^2)^{1/2}/(\psi + \beta\delta)$. For this case equation (6.9) has the following form:

$$Q_t = N(y_t; 0, 1) \tag{6.12}$$

Writing the fraction x_t/n of firms which have adopted the innovation for Q_t, we can use (6.12) to generate y_t and then regress it against time: $y_t = b_0 + b_1 t$. According to equation (6.11), the estimate of b_1 can be equated to $1/\sigma_D^*$. As in equation (6.4), b_1 can also be taken as a measure of the speed (or rate) of diffusion. The expression for $1/\sigma_D^*$ is thus of interest:

$$\frac{1}{\sigma_D^*} = \frac{\psi + \beta\delta}{(\sigma^2 + \beta^2\sigma_s^2)^{1/2}} = \frac{\psi/\beta + \delta}{\{(\sigma/\beta)^2 + \sigma_s\}^{1/2}} \tag{6.13}$$

Of course, a somewhat different expression for $1/\sigma_D^*$ would be obtained if A_t was specified differently. However, in this particular case (in which the diffusion pattern happens to be approximately logistic) there are five parameters identified as influencing, in a manner that is self-evident from (6.13), the rate of innovation diffusion. Of these parameters only δ and σ_s can be measured independently. Davies is also able to estimate σ/β (by regressing y_t, calculated from (6.12) for a given time and a number of size classes, on the logarithm of the average firm size for each class). The estimate is found to be positive for each of the twenty-two innovations. Hence β is also positive, thus providing evidence for the apparently widespread presence of positive scale effects in industrial imitation (see equation (6.8)).

A further step can be made by substituting $\alpha \exp(\psi t)$ for A_t in equation (6.8) and noting that, for firm i of a given industry, $R_{T_i}^*/R_{T_i}^e = 1$ at the time of adoption T_i. If we take $t = 0$ when the invention is first adopted, T_i is also the adoption lag. From (6.8) it follows that

$$T_i = \text{constant} - \frac{\beta}{\psi} \log S_i + u_i \qquad (6.14)$$

where u_i is an error term. In (6.14), β/ψ is the key parameter. Its value can be found from equation (6.13) or, if the adoption lags are already known, by estimating equation (6.14). The values of β/ψ for the sample of innovations studied by Davies were derived from (6.13), and these values imply that, on average, a 1 per cent increase in a firm's size reduces its adoption lag by 0.8 per cent. This rather substantial reduction would then be another manifestation of the presence of the scale effect in imitation.

In most industries the average firm scale has risen very considerably over the last two centuries, possibly by some ten to a hundred times. If the scale effect had been as strong and widespread in the past as Davies's evidence suggests that it is at present, a rise of this magnitude would have been an important contributory factor underlying the apparent long-term tendency for the adoption lags to decline.

Any systematic decline in the average estimation lags adds to the trend rate of technological progress. However, the absolute annual reductions in the imitation lag must be falling, with the lag itself becoming shorter. Therefore the corresponding component of the observed technological change must be gradually withering away. The incentive for increasing firm size should thus also be expected to decline as a consequence of the imitation lag approaching its lower bound, although of course other possible incentives for firm growth may still remain.

The game-theoretic approach: a model by Grindley

A microtheorist may be justifiably uneasy about the models of diffusion, such as those discussed above, in which the decision to adopt an innovation by one firm is seen as having no effect on the profitability of the innovation in any other firm. For, in a competitive environment in which firms adopt an innovation at different times, those which adopt first would be expected to increase the market share either by reducing the unit cost and price or by introducing a more attractive product. As a result the other firms may suffer a loss in relative profits. Each firm, whether considering the timing of adopting a known innovation or not, would presumably be aware of these wider consequences and would attempt to take them into account. The proper analytical framework for considering the diffusion process in the presence of such competitive interactions between potential adopters of an innovation – interactions which are spread over a period of time – is a strategic game. An early example of such a game-theoretic approach in the theory of innovation diffusion is that of Reinganum (1981a,b), which was further developed by Fudenberg and Tirole (1985) and Quirmbach (1986). Other models in this family include those of Flaherty (1980b) and Jensen (1982).

An instructive example of this approach is recent work by Peter Grindley (1986, 1988). In this particular model, the principles of which are outlined below, it is also assumed that the resource cost of adopting an innovation declines over time as a result of learning-by-doing effects or specific R&D effort. There may therefore be competitive benefits as well as losses from waiting. In contrast with the model proposed by Reinganum (1981a, b), the learning rate may increase each time a new firm adopts the innovation.

In the Grindley model the innovation is capable of reducing marginal costs, which are assumed constant and firm-specific, by the same amount ε for all N firms. For a linear demand function, the Cournot solution gives equilibrium quantities of output q_k and profits π_k, $k = 1, \ldots, N$. When firm i adopts an innovation, its profit increases by an amount in proportion to firm size q and ε, while the profits of all the other firms fall. Each firm is assumed to know the production costs of all firms, as well as ε and the learning curve. Adoption of the innovation by all firms enables them to increase their combined profits. However, the distribution of the profit gain depends on the sequence of adoption chosen in terms of dates and firms. The key assumption of the model is that competition among firms rules out all sequences except the one(s) which brings the same benefit to each firm. This competition may result in an outcome in which adoption occurs earlier than optimal, and as a consequence firms make lower profits than if they had all waited. Thus this outcome is an example of market competition leading to a faster rate of diffusion than is socially desirable. Since the benefit resulting from adoption is proportional to firm size, larger firms would adopt earlier. Similarly as in Davies's model, the shape of the diffusion curve, in terms of the number of adopters, is strongly influenced by the industry's size distribution. The general requirement for an S-shaped curve is that relative size differences between firms are greater at the lower and upper ends of the distribution. This requirement is met by the usual distributions: uniform, log-normal, normal, and Poisson. If all firms are of the same size, the learning effect would still imply diffusion, as in the epidemic model, but the diffusion curve, in terms of number of innovating firms, would be increasing exponentially with time rather than of the logistic S-shape. With no cumulative learnings, the diffusion curve in this model mirrors the size distribution of firms as in Davies's (or any probit-type) model. The Grindley model is also consistent with the empirical findings indicating that diffusion is faster for more profitable innovations, and is probably faster in more competitive (less concentrated) industries. On the other hand, stronger learning potential in an innovation tends to result in slower diffusion, as it pays firms to wait longer.

In empirical studies of diffusion it is noted that newly created innovations undergo a particularly large number of improvements. However, the diffusion processes are invariably studied as if we are dealing with the same innovations. In the model outlined above, any subsequent improvements of an innovation would increase the cost reduction parameter ε. In such cases this parameter is typically small and rapidly increasing at the beginning of a diffusion process, is fairly large later on, and declines in the long run. According to this model such changes in ε would, of course, reinforce the S-shaped diffusion pattern.

Chapter seven

The behaviour of enterprises and innovation characteristics in centrally managed economies

The paradox of a high inefficiency and (until the late 1970s) respectable innovation rate

An important aspect of the Marxian theory of economic development and institutional change is an idea that economic efficiency, innovation rate, and ultimately productivity levels are key factors that decide, in the course of history, the outcome of the competition between different forms of organization of economic activity. This idea rests simply on seeing societies as constantly searching for ways of achieving the highest standard of living. Systemic changes, whether evolutionary or revolutionary, are assumed to be an outcome of that search. The theory has been and probably continues to be highly influential in shaping the ideological make-up of the political leaders and their economic policies in the USSR, the People's Republic of China, and other centrally managed economies (CMEs).

The USSR in 1917, most of Eastern Europe in 1945, and China in 1948 started from positions of technological inferiority. Therefore the productivity levels in all these countries were initially low. Thus, to prove the superiority of the Soviet-type economic system, it has been (for Eastern Europe no longer is) essential for them to achieve an internationally superior rate of innovation at least until the gap in output per man-hour could be closed.

To this end, the countries concerned, the USSR above all, have placed an extraordinary emphasis on technical education, R&D, and industrial technological innovation. The Soviet R&D sector has been expanding since 1928 at so rapid a rate that, according to a study by Nolting and Feshbach (1979), the number of Soviet R&D scientists and engineers in 1978 was 'nearly 60 per cent greater than the US'. In the 1980s the non-personnel expenditure on R&D was probably of comparable size in the two countries. This remarkable (quantitative but not qualitative) progress has been accomplished despite the unusually high human and material

94

losses that the Soviet economy sustained in the war years 1914–23 and 1941–5, as well as those from the massive government terror, especially in the 1930s. The USSR was also successful in building up, in a short period of time, a vast education sector which now supplies about twice as many technicians, engineers, and scientists as the US sector (but again the qualitative differences are marked).

The pace of industrialization in the USSR has been slower than in Japan, but apparently of comparable speed to that experienced in Western Europe and faster than in the United States. Two good indicators of the relative Soviet position *vis-à-vis* the United States are the Soviet per capita gross domestic product (GDP) and per capita consumption, each expressed in terms of the corresponding US level. According to Bergson (1963: 288) 'Soviet per capita consumption at the outset (1928) probably did not exceed $160. This is in terms of 1929 prices. In the United States, the corresponding figure as early as 1869–78 was about $190.' US real (in 1929 prices) per capita consumption increased from 1869–78 to 1929 by a factor of 3.7, reaching $700 in 1928 (derived from Bergson's time series (Bergson 1961: 284, Table 78; original data provided by S. Kuznets)). Therefore Soviet per capita consumption, in 1928–9, was at most 23 per cent of that of the United States. In 1955, 'in per capita terms, the Russians are now producing about 40 per cent as much goods and services as the United States' (Bergson 1961: 297). Therefore, assuming that consumption represents about 80 per cent of US GDP and about 60 per cent of Soviet GDP, Soviet per capita consumption in 1955 was about 32 per cent of that of the United States. In 1976, 'in dollar prices, the USSR produced final goods and services equal to 74 per cent of the US national product' (Hughes and Noren 1979: 377). Thus in 1976, in per capita terms, the USSR was producing about 60 per cent as much as and consuming about 48 per cent as much as the United States. These data imply that, during the period 1928–76, the USSR reduced its distance from the United States from about 70 years to about 35 years with reference to per capita consumption, and from about 70 years to about 26 years with reference to per capita GDP. Qualitatively similar results are implied by US–Soviet comparisons of per capita national income in rouble prices. According to one Russian source, that type of income in the Russian Empire in 1913, at 114.3 roubles (1913 value), was 16.8 per cent of the corresponding US level (Koniunk-turnyi Instytut 1926: 12, Table 1). In 1976, 'in ruble prices, Soviet GNP was 50 per cent of US GNP' (Hughes and Noren 1979: 377). This translates to 40 per cent on a per capita basis. Thus in rouble terms, the relative per capita income (and consumption) gap that still remains is significantly larger than when dollar prices are used. Nevertheless, also in these terms, a considerable catching-up has apparently taken place.

The Hughes–Noren and other Western estimates of Soviet national product tend not to reflect adequately qualitative differences between Soviet and US goods and services, thus overestimating Soviet per capita output, especially consumption. However, if static (allocative and X) inefficiency in the USSR is higher than that in the United States, which seems certain, then the technological gap should be lower than the gap in terms of per capita output. Moreover, if it is assumed that the relative static inefficiency gap has remained constant from 1955 to 1976, a rise in per capita output during this period from 40 to 60 per cent of that of the United States in dollar prices (and from some 28 to 40 per cent in rouble prices) must reflect a fall in the economy-wide technological gap. The latter conclusion would be strengthened if, compared with the US economy, the Soviet economies of scale were less and static inefficiency increased more during that period, and it would be weakened if the Soviet capital-to-output ratio had been increasing in relation to the US capital-to-output ratio. The outputs of the two key industrial sectors, one producing machinery and equipment for investment purposes and the other defence goods, are also thought to have outpaced the corresponding outputs of the US industries. Productivity performance has not been poor either. In Soviet industry, the trend rate of growth of output per man-hour in the years 1928–75 was about 5.5 per cent according to Soviet data and some 4 per cent according to Western estimates. This is lower than the equivalent growth rate for Japan, but comparable with that for Western Europe and higher than the rates in the United States and the United Kingdom in the same period.

However, a large body of microeconomic evidence has been accumulated in Eastern Europe and the USSR which indicates a high degree of resource misallocation in both conventional production and R&D, a large but rather slow and often wasteful amount of investment activity, and a generally high resistance to innovation, especially in existing enterprises. Despite its large size, equal to at least a quarter of the size of world R&D activity, the contribution of Soviet and East European R&D activity to the world flow of new inventions is negligible. In the 1970s, member countries of the Council of Mutual Economic Assistance (CMEA) were importing about ten times more licences in terms of dollars paid than they exported; the exports represented merely 1 per cent of the estimated total of world exports. The analogous ratio for

Western Europe in recent years has been approximately 2, and for the United States about 0.2. Similarly insignificant is the Soviet and East European share of the Western world market for manufactured products, with these countries continuing to exchange mainly raw materials and standard intermediate goods for Western imports. Another indicator of poor efficiency is the apparently excessive use of primary and intermediate inputs per unit of final demand output. A high energy intensity in CMEA countries, measured in terms of the consumption of energy per unit of GNP, has been noted by many authors. Slama (1986) estimates it to be, on average, 2.3 times higher than the energy intensity of the European OECD countries. Moreover, following the 1973 energy price rise, this unit consumption has been falling more in the West than in the CMEA as a region (Gomulka and Rostowski 1988). Gomulka and Rostowski also estimate the material intensity in Czechoslovakia, Hungary, and Poland in terms of the value of intermediate inputs per unit of gross value added. The measure is found to vary significantly among the twelve OECD countries which were used for comparative purposes. In terms of arithmetic averages, the three CMEA countries in the years around 1975 were about 30 per cent more material intensive (the confidence interval at 95 per cent probability level ranged from 24 to 61 per cent). If energy prices are adjusted to international levels, the excess of material intensity has increased to about 50 per cent. Since the adoption of the OECD's composition of final demand by the CMEA's three countries was found to leave their material intensity almost unchanged, the principal cause of the high intensity is thought to be the economic system itself, meaning above all the incentives, the prices, the decision criteria, and the ways that the decisions are arrived at and implemented.

We have thus outlined what seems to be an important paradox of CMEs: an apparently low static economic efficiency and yet a respectable rate of change in the (joint) input productivity, indicating scale economies of a high dynamic efficiency. An interpretation of this Schumpeterian-type paradox will be offered in Chapters 11 and 12. In what follows we shall bring together some of the major pieces of evidence of the microeconomic type with a view to identifying the systemic characteristics that influence the choice of technique and the direction and rate of the innovation activity in CMEs.

More specifically, this chapter is structured around the following five topics: (1) the differences in the environment in which socialist and capitalist enterprises operate; (2) the effect of such systemic differences on enterprise motivation and technical/

production choices; (3) the effect of these systemic differences on enterprise innovation behaviour; (4) the extent to which the Hungarian New Economic Mechanism, which in the 1980s was adopted by China, Poland, and to a lesser extent the USSR, differs from the more traditional Soviet-type economies in these respects; (5) the arguments for a much higher degree of marketization and competitiveness to improve innovation performance, together with grounds for doubting whether such an improvement can be achieved without large losses with respect to key socialist aims such as low income and wealth inequality, high employment security, and negligible private economic activity.

As far as (3) is concerned, we shall identify six aspects that appear to be particularly characteristic of the R&D and innovation activity of the enterprises under Soviet-type socialism. These are as follows: (i) the main innovation drive comes from the centre; (ii) the decision-making freedom and own resources available to enterprises in implementing inventions are severely limited; (iii) supply difficulties are an important factor in innovation – sometimes as stimulus, sometimes as constraint; (iv) enterprises do little on their own not only because they cannot do much but also because they do not need to innovate in order to survive; (v) despite innovation pressures from above, quantity is emphasized at the expense of choice and quality; (vi) the time-lag between innovating and inventing is higher and innovation diffusion slower than in the developed market economies. For Eastern Europe of the early 1990s, most of what follows is already history. However, for the USSR and China much of the old system is still the reality.

The discipline of the plan and the freedom of the firm

We shall begin by contrasting a typical enterprise in a CME of the Soviet and Chinese variety with a market economy firm of the capitalist or mixed-economy type operating in a fairly competitive market environment. We shall also note the implications of the recent reforms in CMEs, especially in Hungary, China, and the USSR, for the behaviour of the enterprise.

The survival and growth prospects of the capitalist firm depend, of course, on its profitability performance. This performance is in turn influenced by prices, which are being established in the input, financial, and output markets where our representative firm meets its competitors and bids for resources and customers. Thus, to the ____t is a price-taker and receives no subsidy, the capitalist ____cted to the financial discipline of the market.

____rprise in a CME is requested to keep its own accounts

and allowed to operate formally as an independent unit for managerial and supervision convenience only. By definition, it is merely a subsidiary of limited financial and managerial responsibility of a larger organizational unit, ultimately of the entire (centrally managed) economy. It operates within an organizational structure in which enterprises producing similar goods would typically form a production association, several associations would form a branch, headed by a ministry, and the activities of the ministries would be directed and supervised by the central (political and economic) authorities, in particular the politburo, the prime minister, and the central planning commission. Leaving aside the details of the spheres of responsibility and the operation of the various levels of the decision-making hierarchy, it will be useful for our purposes to identify and discuss the situation of the enterprise in terms of the following: the key constraints which define the firm's production possibility set, the major concerns that motivate it to make particular production/technological choices, and the forms of innovation (and other) activity that, in theory at least, are aimed at lifting the binding constraints.

Using the same notation as in Chapter 1, let x_j represent the scale on which production process j, where $j = 1, \ldots, m$ can be operated and let $b_j = \{b_{ij}\}$, $i = 1, \ldots, n$, denote a (column) vector of the n types of inputs used and outputs produced when the process j is operated on a unit scale. By convention, $b_{ij} < 0$ if good i is actually an input (or negative output) in process j, $b_{ij} > 0$ if good i is actually an output, and $b_{ij} = 0$ if i is neither used nor produced. In general, the input–output coefficients $\{b_{ij}\}$ may depend on x_j, and the x_j themselves may either be continuous or take on certain discrete values. When the m processes are operated on the scales $x_1, x_2, \ldots, x_m$, the net outputs produced are given by the (column) vector

$$BX = \left\{ \sum_{j=1}^{m} b_{ij} x_j \right\}$$

Furthermore, let $\bar{Z}$ be a (column) vector of minimum output targets, entering with a positive sign, and maximum input quotas, entering with a negative sign, which the enterprise should meet and may employ respectively.

The first major difference between a market economy firm and a CME enterprise is in the specification of the vector $\bar{Z}$ (Figures 7.1 and 7.2). In a market economy, capitalist as well as socialist, the minimum outputs are usually low as, with the exception of long-term contracts, there is typically no externally imposed

99

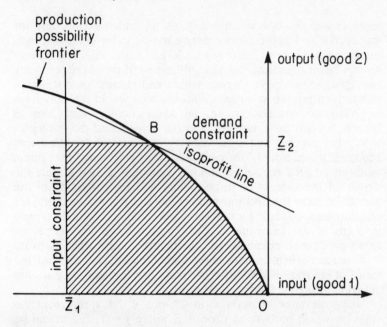

Figure 7.1 The capitalist firm: at the (profit-maximizing) point B the demand constraint is binding. The shaded area is the set of feasible techniques.

requirement to produce anything, while the maximum inputs could be very large since, with the exception of fixed capital and some labour skills in the short term, most inputs can be obtained easily in domestic or international markets.

In contrast, our representative CME enterprise would negotiate with the higher (planning and managing) authorities a production plan that would specify the minimum output targets and the maximum input allocations, almost always with the specific recognition that the negotiating sides have conflicting interests in the outcome. Negotiations of this type cannot and do not specify the vector $\bar{Z}$ fully. Since the total number of different (types of) goods produced in any modern economy runs into tens of millions, the central planning and management authorities select from among them a subgroup of key goods called centrally (as against locally) planned commodities. For each of these commodities the central planners draw up the so-called product or material balance, attempting to bring demands and supplies into equilibrium at the national level. The number of these product balances is small, some 200–2,000 at the central planning commission level. Most of

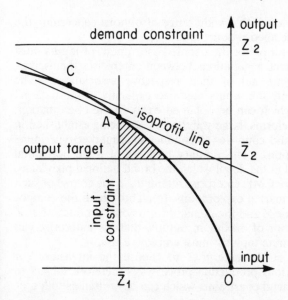

Figure 7.2 The socialist firm: output and resource constraints reduce the allocation role of prices. At the (profit-maximizing) point A, the resource constraint is binding. At the same prices, point C would be chosen in the absence of the resource constraint.

these 'products' are therefore not individual product types but composite products, or groups of similar goods, usually expressed in both physical terms (e.g. square metres of wool products, tons of rolled steel, or the number of radio sets) and value terms. These targets are addressed primarily to ministries which will disaggregate them considerably. The number of output targets that reach the association level varies from a few thousand for most CMEs to about 40,000 for the Soviet Union (before the reforms of the late 1980s). Each such target is, typically, still an aggregate of several to several thousand individually priced goods. The input quotas are similarly aggregated. The association is not only an intermediate level representing enterprises *vis-à-vis* ministries and ministries *vis-à-vis* enterprises, but is also a meeting place where the distribution of the association ouptut targets and input quotas among its members is negotiated and decided (the ministry is such a meeting place for the associations, and the central planning commission for the ministries). Since these targets and quotas are typically large aggregates, the plan can be viewed as representing a set of (resource, demand, and financial) constraints such that the enterprise is still left with a wide range of options concerning the selection of the detailed product mix.

However, the supply of the detailed assortment of inputs must be secured in direct negotiations between enterprises themselves, sometimes with the aid of their respective associations. (The exceptions are only the inputs specified explicitly in the national plan and those which can be obtained easily in the open market. Under the new reforms, large warehouses are being established in some industries and countries to conduct such market-type inter-enterprise operations.) The primary business of these horizontal links is thus to fill in the detail which the broad national plan lacks. Our representative CME enterprise attempts, in the course of such negotiations and market explorations, to identify both the components of the vector $\bar{Z}$ and the demand for its own products. Let $\underline{Z}$ represent the vector of maximum outputs that the enterprise can sell and the minimum inputs it must employ or use.

The freedom of the enterprise to choose the intensities (or scales) of the various production processes is therefore subject to resource and demand constraints which can be written as follows:

$$BX \geqslant \bar{Z} \qquad BX \leqslant \underline{Z} \qquad\qquad (7.1)$$

where $X = \{x_j\}$, $j = 1, \ldots, m$, is the (column) vector of process intensities.

The choice of X is further constrained by what can be called the financial constraint. For the capitalist firm this would be the maximum financial loss that the firm can sustain. The CME enterprise usually also has financial targets to meet. These targets relate to a number of different categories of costs, as well as to overall profitability and/or the rate of profit. However, given widespread price anomalies, these targets would be the subject of bargaining between the firms and higher levels of decision-making. This gives the firm ability to co-determine its income through ways other than its own production and innovation effort. To use Kornai's term, the financial constraint is softer under socialism than capitalism (Kornai 1980). (See also the exchange between Kornai and Gomulka in *Economics of Planning* (1985) concerning the meaning of the concept and its implications for efficiency and shortages. The central question of this exchange is whether shortage and inefficiency – the major features of Soviet type economies – are necessary and unavoidable consequences of budget softness. Kornai's answer to both, particularly to shortages, is in the affirmative, whereas Gomulka argues that market equilibria can be sustained and only inefficiency is essentially unavoidable.)

All the imposed targets, some of them fairly aggregated, form the directive plan. This plan represents a set of constraints for the choice of the enterprise's own plan in the form of specific quantities of inputs and outputs, and their financial implications.

If the enterprise operates its production processes at the intensities X, then BX is the vector of net outputs (negative net output indicates that the good is actually an input). Let P stand for a vector of prices. Hence, PBX is the difference between the value of all (actual) outputs and the value of all (actual) inputs, thus representing pure profits. The workers and managers of the CME enterprise would not normally aim to maximize profits. Instead, a performance-related part of their total income, which can be as high as 50 per cent or more for top managers and key workers in priority areas of production and investment, represents the sum of bonuses related to the fulfilment of the various targets. Denoting this sum by B^{total}, we have

$$B^{\text{total}} = \sum_k f_k\{a_k - t_k(x)\} \tag{7.2}$$

where $a_k - t_k(x)$ is the difference between the value t_k of target k and the achieved level a_k, where the latter depends on the vector of intensities to be chosen by the enterprise. The choice of its own plan by the enterprise thus depends on the particular choice of targets by the higher levels and the bonus weights associated with the fulfilment of these targets. Although there would usually be a large number of physical and value targets that earn material bonuses (as well as medals and promotions), the Soviet planners originally chose the volume of gross output as the key target. This choice has proved to have severe economic drawbacks. When the output target is expressed in physical terms only, it encourages enterprises to use technological processes which substitute as much quantity for quality as is possible within a permissible quality range. When many goods are produced and the output target is expressed in value terms, the output prices immediately become important. These prices are typically established to reflect average industrial costs. It follows that the producers would have an interest in selecting products and processes which require the most expensive material inputs to produce expensive goods, thus requiring little effort to fulfil the output plan. Although the producer may have to use cheap inputs because of a shortage of expensive ones, in bargaining situations with the higher authorities and suppliers an enterprise would be induced to press for the supply of inputs and the imposition of output targets that would enable it to produce material-intensive products that easily fulfilled the plan.

Since the late 1960s, the global output indicator has been

supplemented with, or replaced by, a related indicator of the value of sales in the USSR and by value added in most of Eastern Europe. Under the reforms of the 1980s the profitability indicator has been substantially upgraded in Poland and China as well as in Hungary. The USSR has been adopting similar reforms since the late 1980s. The changes are intended to strengthen the bargaining position of the buyer, reduce the bias towards material intensiveness in the choice of production processes, increase responsiveness to price changes, stimulate cost-reducing and demand-expanding innovation, and so forth. However, in the USSR at least, central planning and management are to continue to play important roles.

It is legitimate to ask whether the social interest would not be served better if a weighted multitarget pay-off function like (7.2) was replaced by a single bonus motivating the enterprise to maximize, subject to constraints (7.1), a synthetic index of performance such as the size of pure profits or the rate of profit. However, if a directive plan is to have any meaning at all, it should not be ignored, and the problem that central planners must always keep in mind is one of ensuring that the enterprise actually does its best to respect the constraints implied by the plan, in particular the minimum output targets. However, if profits are its only concern, the enterprise may have an interest in ignoring many components of the directive plan. It may also have an opportunity to do so at a low risk since in the uncertain environment in which the enterprise usually operates there may often be legitimate reasons for failing to meet all or some of the imposed targets, thus presenting the planners with difficulty in distinguishing between such legitimate reasons and invented ones. It follows that a bonus system of awards and penalties, such as (7.2), is essential in a CME in order to motivate enterprises which are subject to central planning to respect the constraints specified by their directive plans. A reform which aims to abolish the former must also abolish the latter.

Another problem the central planners face relates to the allocative role of prices. Most product prices in CPEs are state imposed, and for reasons of administrative convenience or necessity tend to be stable over time and highly inflexible, responding little to market disequilibria. Consequently, they are likely to differ substantially from the marginal social costs of production. However, such *ad hoc* prices would play a major allocative role if the choice set of the socialist firm was allowed to be large and the firm was motivated to maximize a monetary (synthetic) target such as pure profits. To reduce that role, Soviet planners used to set high minimum output targets and low maximum input quotas for large numbers of goods, so that the number of all feasible choices open to the socialist firm was small. This practice of 'taut planning', in conjunction with the system of financial incentives and controls, is thus at the core of the centralized system of managing the economy. This system brings to the fore the allocative role of quantities, in the form of resource constraints and fairly aggregate output targets, rather than prices and financial targets.

Aggregated nationally, the differences between the total supplies of and the demands for particular products would thus be a function of the following parameters: prices, bonuses, and all the (resource, demand, financial, and other) constraints that firms face, including those listed in the detailed enterprise plans. In principle, the central planners may adjust the values of these parameters to ensure an approximate equality between marginal social products among firms for the same and every resource type (efficiency requirements) and between economy-wide quantities supplied and demanded (consistency requirement). The difficulty that the planners are up against is that the correspondences between these values and both marginal social products and excess demands are effectively unknown, partly because of the rapidly changing environment (technology, consumer tastes, etc.) in which the enterprises and planners have to operate and act. Moreover, for administrative reasons, not only product prices and bonus criteria must be and have been relatively sticky, but also the number of output targets and input quotas must be relatively small, with the latter implying a high degree of aggregation of the (imposed) directive plans. This high degree of aggregation allows more freedom of choice for the enterprise. The greater freedom could be used by the firm to serve its own interest which, given the simplistic incentive system, is likely to be at variance with the social interest, even one formulated by the central planners. In particular, the marginal social product of the same input is likely to vary widely among firms, and the mismatch between quantities supplied and demanded in markets for particular inputs and outputs is likely to be widespread, high, and persistent.

Shortages and surpluses for tradeable goods can be reduced by imports and exports respectively. However, the imbalances for non-tradeable goods, such as the services of labour or infrastructure, cannot be reduced in this way. Moreover, the combination of taut planning and bonuses that emphasize compliance with output plans rather than profitability performance creates a demand for imports whose value tends persistently to outstrip the value of exports. The Soviet central planners respond to this hunger for imports with a policy of strict import controls, executed through rationing the foreign exchange set aside for associations and

105

individual enterprises by administratively imposed limits. Foreign trade still remains an important instrument which the planners can use first to activate resource reserves by reducing resource bottlenecks and second to upgrade demand constraints by exporting surplus outputs with a view to reducing the incidence of forced substitution of surplus goods for those in shortage. However, severe import restrictions prevent these two important aims from being fully achieved.

In summary, a socialist enterprise of the standard Soviet type operates in an economy that tends to be persistently overheated. Demand constraints tend to be of secondary concern relative to resource constraints. Domestic markets are well sheltered from imports, and competitive processes in these markets are absent or low. Consequently, neither outputs nor prices are changed quickly in response to product surpluses and/or shortages. The income of the enterprise workers and managers is a function of the plan, its targets, and its limits, rather than dependent on its cost and profitability performance. However, the latter is suspect anyhow, in view of the *ad hoc* nature of prices. The enterprise is subjected to the discipline of the directive plan rather than the competitive market, and therefore the bargaining processes that determine the targets and limits play a vital role, with the enterprises tending to seek low output targets and high input allocations. In this way the enterprises are in a position to co-determine their incomes, thus making their budget constraint soft. Consequently, the force of standard purely economic principles in allocative decisions is blunted by the fact that the role of the social interactions (visible hand) that take place between the many decision-makers of the large administrative hierarchy is large and the disciplinary influence of market competition (invisible hand) between enterprises is minor.

Gorbachev's reforms of the late 1980s represent an attempt to diffuse management decision-making, enhance the role of markets, and reduce the role of central directive plans. The emerging system is likely to be a variant of the Hungarian New Economic Mechanism, but will probably still retain a large role for directive planning (for more details see Hewett (1988) and Gomulka (1988)).

Systemic characteristics and policy aspects of innovation in centrally managed economies

The subject of R&D and innovative activity in the USSR and Eastern Europe – its organization, size, and performance – has been studied particularly intensively since the mid-1970s. Among the most systematic and substantial Western writings in this field are those of Zaleski *et al.* (1969), on the organization of R&D and science policy, and Berliner (1976) on innovative activity itself. A concise and incisive survey of this by now extensive literature, Western and Eastern, has been made recently by Hanson and Pavitt (1987). In this chapter we shall be rather selective in the choice of institutional detail, emphasizing instead what appear to be the major systemic and policy features of innovative activity under central planning.

The approach is comparative. In the first instance we shall contrast the innovation process in traditional CMEs with that in market-based developed Western economies. The primary purpose of this comparison is to identify the differences. The major deficiencies of the innovation process in CMEs will become apparent, and we shall note the economic effects of these deficiencies. Second, we shall briefly discuss why the innovation effects of reforms in Hungary, Yugoslavia, and Poland have, until the late 1980s, proved to be insignificant.

Major systemic characteristics of the innovation process

(i) *The main innovation drive comes from the centre*: since innovation is usually linked with investment in new plant and machinery, and since most investment resources are centrally controlled, it is the offices of central institutions where decisions on innovations, especially those concerning major projects, are made. These central decisions are made typically on grounds other than cost–benefit analysis. New major products and processes are identified by the technical staff (including R&D staff) of these institutions. These innovations are typically already in existence abroad, and resources are deployed to import or to imitate them. Such major innovations lead to a sequence of other innovations, in supplying industries, to produce the components or ingredients of required standards of quality and performance. New enterprises are often set up to implement such major innovations.

A corollary of this feature, and this is our second characteristic, is that (ii) *the decision-making freedom and the resources available to enterprises for implementing innovations which originate at the enterprise level are very limited.* Moreover, the financial incentive for undertaking an innovation for the enterprise personnel is also weak. Inventors are expected to be motivated primarily by professional ambition rather than financial benefit or promotion prospects. This feature is not really surprising, given the limited reliance of the system on entrepreneurial activity or enterprise.

Nevertheless, many small and medium innovations are adopted by enterprises. However, and this is our next characteristic, (iii) *innovating firms are often motivated by the need to overcome supply difficulties.* The need to adapt to new demand conditions, especially in export markets, has also been found to be important, but enterprises are resource rather than demand constrained and as a rule take little initiative in exploiting new technological opportunities for the purpose of creating new demands at home or abroad, something that often happens in market-oriented economies. *The enterprise's innovation strategy would thus appear to be primarily defensive rather than offensive,* with the resource-constrained rather than the demand-oriented nature of the economy emerging as a key underlying factor. An implication of this third characteristic is that process rather than product innovations tend to dominate. Product innovations serve to sustain or stimulate demand. However, demand tends to be excessive in CMEs, whereas supply difficulties are common. To cope with them firms would use the same tools and equipment but investigate the use of somewhat different intermediate inputs (ingredients) to produce essentially the same product. Consequently, many of the new process innovations reflect the widespread phenomenon of forced substitution, and as such represent technological necessity rather than actual improvement. One study reported that, in the judgement of firms themselves, 30 per cent of the innovations in the sample represented technological regress and that, in most cases, no serious cost–benefit analysis was undertaken (Kubielas 1980).

The soft budget constraint of R&D organizations may not harm, and sometimes may even help, the researchers. However, the important implication of both the soft budget constraint of enterprises themselves and the absence of competition is that the threat to any enterprise of elimination from the market is virtually absent. This gives rise to the presence of an apparently wide gap between considerable overall supply of and limited enterprise demand for the invention–innovation activity. Hence, in terms of their own innovative activity, (iv) *enterprises do little on their own not only because they cannot do much but also because they do not need to innovate in order to survive* (nor do they gain much if they can and do innovate; see characteristic (ii)).

(v) The fifth characteristic is that *enterprises in CMEs tend to be large in scale and to trade off choice and quality for quantity.* Market-oriented economies are characterized by the presence of large risk capital, the purpose of which is to sustain a high rate of birth of enterprises set up to exploit promising domestic or foreign inventions. The innovation/investment decision is thus diffused and decentralized, and these numerous small-scale enterprises are often used as a testing ground for new inventions. The market is thus used as a screening device for the purpose of channelling resources from old to new industries in a rational way. As mentioned earlier, in CMEs screening of almost all innovation possibilities which involve significant investment expenditures is centralized and risk-taking is almost fully nationalized. The limited screening capacity of the centre would call for the construction of new firms that are limited in number and large in scale. Such a bias for large size in CMEs has indeed been well documented by studies of size distributions of firms in the two types of economies. If scale economies are present, as they often are in the manufacturing sector, then suppressing choice and emphasizing scale is also an effective method of raising output levels. For example, summarizing his survey of a major industry, James Grant (1979) notes:

> The Soviet machine tool industry, developing independently of western assistance, has become the world's largest producer of machine tools. However, emphasis has been on large-scale production of relatively simple-to-produce, general-purpose machine tools at the expense of special-purpose and complex types.

(vi) *In CMEs the time-lag between domestic inventing and innovating is high and the subsequent spread of inventions tends to be slow.* In terms of the two measures of innovative responsiveness – the average time-lag between (domestic) inventing and innovating and the speed with which an innovation (domestic or foreign) subsequently spreads through an economy – the CMEs appear to lag significantly behind the major Western economies. To give an example, a comparative study by Martens and Young (1979) finds that for a fairly large sample 'the US and West Germany implemented over 50 per cent of their [domestic] inventions in little more than one year, whereas the Soviets needed slightly more than three years to achieve this percentage of implementation'. The rate of diffusion of new steel-making technology has also been markedly slower in the USSR than in the West. The first oxygen converters were introduced in the USSR, the United States, the FRG, and Japan in the period 1954–7; by 1974, oxygen steel as a proportion of total steel output was 23 per cent in the USSR, compared with 56 per cent in the United States, 69 per cent in the FRG, and 81 per cent in Japan (Figure 7.3). It is worth noting that the steel-making industry is a relatively modern and high-priority

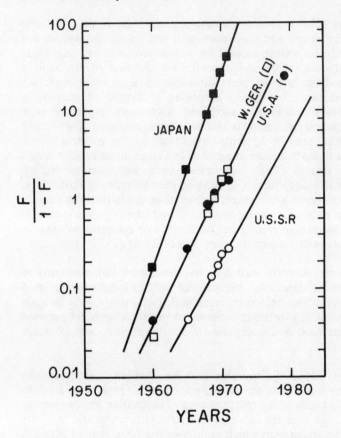

Figure 7.3 Substitution of oxygen steelmaking for the open hearth and Bessemer methods in Japan, USSR, Federal Republic of Germany and the United States since 1960. (*F* is the extent of substitution on a 0 to 1 scale.)

Source: Pry, R.H. (1974), in H.F. Davidson *et al.* (eds) *Technology Transfer*, Leiden: Nordhoff, p. 232

Soviet industry, and that diffusion was slower despite a faster rate of capacity expansion than in the United States, suggesting that the example may well be a good indicator of generally slower rates of diffusion throughout Soviet industry.

The negative implications of these innovation characteristics are fairly powerful and relatively well documented. As mentioned earlier, despite its large size, the contribution of Soviet and East

European R&D activity to the world flow of new inventions is negligible. In terms of invention patents granted to nationals abroad, CMEs account for about 3 per cent of the world total (Chapter 10, Table 10.2). In terms of exports of licenses, CMEs account for about 1 per cent of the world total. In both categories developed market economies account for about 96 per cent. In terms of manufacturing exports, the USSR and Eastern Europe account for about 2 per cent of the world market. Perhaps even more importantly, the home consumer is offered goods and services of poor quality and narrow range. Moreover, processes which are relatively highly material intensive are used to produce these goods. This has become particularly evident during the 1970s and 1980s when market economies responded strongly to much higher prices of energy and other raw materials whereas CMEs did not.

These negative implications of the centralized system were also present in the 1930s and in the 1950s when the productivity gains in Soviet industry were rather high, indicating that the innovation rate was also high. An explanation of this paradoxical phenomenon, to be developed further in Chapter 11, rests on the fact that at a low level of development a substantial part of industry is highly backward and a small part is fairly modern. Consequently, an internal diffusion of technology and skills within industry (the economy) as a whole takes place as the modern sector expands and the backward sector shrinks. The rate of this diffusion is related to the size of the investment effort in the modern sector (and the rate of expansion of the education sector). This strategy is what can be called an 'extensive way of intensive growth'. However, the strategy's effectiveness ends once the backward sector disappears. At that point the innovation rate falls to that observed in the world's technology frontier area. An innovation slow-down of this kind is therefore a mark of success, not failure. However, the success is only partial if the modern sector of a country lags significantly behind the world's modern sector. This is in fact what we observe with respect to CMEs (Chapter 9, third section). This lag is linked to the higher time-lags in the invention–innovation–imitation–diffusion sequence. Thus the systemic features of the innovation process ultimately have the effect that CMEs will permanently have a much lower standard of living – a third or so of the level observed in the most developed market economies – unless radical reforms are undertaken.

The effects of Hungarian-type reforms

The small East European countries have shared with the USSR the main weaknesses of its innovative activity, without enjoying some of its strengths such as a large R&D sector, the possibility of exploiting economies of scale, and presumably considerable competitive pressures in military-related research and innovation. These small countries are not without advantages of their own, however, the main one being that they trade relatively more with the developed West. In particular, the share of machinery imports in total machinery investment in Eastern Europe, at about 10–30 per cent, is roughly five times greater than the share of imports in Soviet investment. The presumably much larger technological transfer from the West compensates to some extent for their much smaller (original and imitative) R&D activity. The greater exposure to Western markets for manufactured goods has also produced some competitive pressures of its own. In order to withstand these pressures, the East European governments have been forced to think seriously about introducing systemic changes that would give greater managerial and financial autonomy to the exporting firms. In the case of two countries, Hungary and Poland, major new changes of this kind have actually been undertaken since late 1989 and early 1990. The earlier Polish reform of 1981–2 was modelled on the Hungarian New Economic Mechanism, which was introduced in January 1968, re-introduced in 1982, and reformed in the late 1980s. The Chinese and Soviet reforms of the early 1980s, while substantial, are no more radical than the Hungarian reform, all being somewhat different variants of essentially the same system (for more details see Kornai (1985) and Gomulka (1986b, 1987)).

Under this system, prices in Hungary and Poland were already much more flexible and rational, enterprises nominally had no centrally imposed output targets and input quotas, and the role of the profitability principle in firms' own choices was significantly larger. Since plan fulfilment is no longer the main criterion of management evaluation, the negative economic consequences of the practice of bargaining for low output targets and high input quotas were less in evidence. Yet most of the major investment and price decisions and key managerial appointments continued to be made centrally, and they were used for the purposes of influencing the production, market, and innovation decisions of enterprises, especially large enterprises. As the local currency was not convertible, domestic markets remained well sheltered from foreign competition. A large number of these markets continue to be dominated by one firm. The absence of bankruptcies was a further indication that financial constraints have remained soft and market competition low. Moreover, profits are taxed at a high rate, in Poland on average at 60 per cent. Consequently, highly successful enterprises are prevented from expanding rapidly, while loss-makers have until recently been prevented from closing down.

There was nevertheless evidence of some improvement under the NEP. For instance, monetary incentives for inventors have been much strengthened. Enterprises have more investment resources to play with and are therefore in a stronger position to initiate innovation. In Poland, it has been possible since 1987 to form financially independent innovation enterprises. These enterprises sell inventions at market-determined prices, can retain half their dollar earnings, can offer high incomes to inventors, and are free to attract private capital, domestic or foreign (*Dziennik Ustaw PRL*, 28 July 1987).

However, the key problem remains. This is that enterprises are still not particularly keen to innovate: *they continue either not to gain much by inventing or not to lose much by not inventing*. The old pressure to innovate – that of the centre – is much weaker than it was before, but the new pressure to innovate – that originating from the competitive market – is still weak. In terms of our usual indicators, such as patents implemented, productivity improvements, and manufacturing exports, the macroeconomic evidence is one of deterioration rather than improvement. Until 1990, authorities in Hungary and Poland used tax concessions and outright subsidies to induce innovation, particularly cost-reducing process innovation and export-augmenting product innovation. However, the concessions involved are relatively small. The major fact remains that wages and wage increases are almost independent of the efficiency indicators of enterprises (Kornai 1987; Schaffer 1989). This is in part related to the fact that the authorities have not developed any effective way of controlling wage inflation other than through administratively imposed wage controls or highly progressive wage taxes. However, these fiscal instruments tend to weaken the incentive to work harder and to innovate faster than the average worker or the average enterprise (Gomulka 1986b: Chapter 14). Here, as well as in other key areas of the new reformed system, strong limits to innovation still appear to be at work, the roots of which can be traced to policies intended to achieve standard socialist ends, such as high job security, low income differentiation, and limited private ownership, features which only the new post-communist authorities in these countries wish to change.

Part two

Macroeconomics of innovation, technology transfer, and growth

Innovation biases, factor substitution, and the measurement of technological change: definitions and theory

In this chapter we begin with a discussion of the following question: how do we distinguish a change in an input–output combination made by a firm (or economy) in response to a movement in relative prices from a change in this combination made in response to an innovation? The aspects to be considered are theoretical and empirical. We first define measures of these two responses, to be called the substitution effect and the technological bias effect respectively. We then note that, at least in theory, a separation of the two effects is possible, and we discuss how this can be done for several alternative ways of defining the effects. We shall pay particular attention to two such definitions, one due to Hicks and another to Harrod, which are both widely used by economists. These two definitions are introduced in the first section, where their many interpretations and mutual relations are also discussed.

Another problem which the economist must somehow deal with, or at least be aware of, is that the data we have are typically insufficient to allow the substitution and technological bias effects, however defined, to be identified and separated from each other. This serious empirical difficulty is discussed in the second section.

One way of partially meeting the separation problem is to deduce theoretically the bias of technological change that would be consistent with the overall pattern of economic growth that we think that we observe over long periods of time. This pattern is often referred to as the Kaldorian stylized facts, and will be discussed in Chapter 9. It suggests that exponential growth, at least in the technologically most advanced part of the world, is a good first approximation of the real world. In the third section of this chapter we discuss the type of technological progress necessary to permit such growth. The outcome of this theoretical discussion is that the observed pattern of growth is as if the technological change has been biased (in the Hicks sense) in the direction of

117

saving primarily the non-reproducible inputs, and perhaps the labour inputs above all.

Not only the bias but also the rate of technological change can be defined in a number of different ways. Their different interpretations and implications are discussed in the fourth section. In particular, we extend to the n-sector case the discussion from Chapter 1 of the measurement of the contribution of technological change to economic growth over long periods of time. In this context we return to considering the so-called reproducibility problem. We also introduce and interpret the rules for deriving aggregate economy-wide rates of technological change if the rates pertaining to sectors are known.

The chapter ends with an exposition of two theories of technological bias. The theories are quite different, yet contain ideas which, it will be argued, are complementary rather than contradictory. These ideas are integrated into a view designed to rationalize the apparent labour-saving bias in technological change. The exposition represents a continuation of the theoretical discussion which begins in the third section.

Innovation biases: two-factor and n-factor analysis

As has been mentioned above, there is more than one way of defining the innovation rate and bias. To begin with, let us assume that output Y is related to inputs of capital services and labour services – assumed to be proportional to the stock of capital K and employment L respectively – through a standard (neoclassical) constant-returns-to-scale production function, so that

$$Y = F(K, L, t) = Lf(k, t) = Lh(v, t) \qquad (8.1)$$

where $k = K/L$ and $v = K/Y$.

Equation (8.1) needs to be interpreted with care. Taken literally, it suggests that technological change is assumed to be completely independent of capital accumulation and hence that it is purely 'disembodied' in type. It also suggests that (dis)economies of scale are excluded. Both assumptions would limit innovation to little more than organizational change and effectively render (8.1) unrealistic to the point of being irrelevant.

In fact it is now widely accepted that the variation between industries in labour productivity levels and growth rates 'can be explained primarily by the uneven impact of three influences: (i) improvements in technical knowledge, (ii) potential economies of scale and the extent of their realization, and (iii) factor substitution', and that 'although analytically distinct, these three influences are highly inter related' (Salter 1966: 143). Indeed, hardly any innovation can be adopted without at the same time altering the input composition in the adopting firm. Also, potential economies of scale are typically created by the discovery of new processes and the invention of new equipment, so that we would usually be in error to view such scale economies as separable from technological change and factor substitution.

Our interpretation of functional relationships such as (8.1) has already been indicated in Chapter 1. Whatever the production unit – whether individual firm, industrial sector, or the economy as a whole – relationship (8.1) must involve a (lesser or greater) degree of aggregation. Thus, as explained in that chapter, in reality a set of production processes, each involving many kinds of inputs and outputs, corresponds to each point of the production function, to be represented in (8.1) by a triple (Y, K, L). In order to aggregate these inputs and outputs, we would have to use prices, probably market prices at a given time. Given these prices, which for aggregation purposes need to be held constant from one period to the next, technological innovations in any period may involve a change in the composition of K, L, and Y and yet be compatible with keeping any or all of these aggregates unaltered.

The argument for the assumption of constant returns to scale is somewhat different. Of course, whatever the aggregation level, returns to scale are unlikely to be constant for all quantities of output. What we (implicitly) assume is that there would always be a range of outputs where, at the level of firms, the returns are in fact constant, and that the (cost-minimizing) firms would operate in that range in every period. This assumption is not necessarily very restrictive if our interest in the relation between inputs and outputs is limited, as is usually the case, only to the neighbourhood of the processes that are actually operated. Technological innovations may move the range in question from one period to another, thus giving rise to dynamic economies of scale. These potential economies are assumed to be captured in (8.1) by the technological progress variable.

Relation (8.1) can be expressed in terms of growth rates. We have, on taking logarithms and differentiating fully with respect to time,

$$\mathring{Y} = \pi \mathring{K} + (1 - \pi)\mathring{L} + \lambda = \mathring{L} + \pi \mathring{k} + \lambda = \mathring{L} + q\mathring{v} + \alpha$$

$$(8.2)$$

where the (full) time derivative of a function $x(t)$ is denoted by $\dot{x}$ and its growth rate by $\mathring{x} = \dot{x}/x$. The notation used in (8.2) is as follows: $k = K/L$, $v = K/Y$, $y = Y/L$, and

$$\pi = \frac{F_K K}{Y} \qquad \lambda = \frac{F_t}{Y} = \frac{f_t}{Y} \tag{8.3}$$

$$q = \frac{h_v v}{y} \qquad \alpha = \frac{h_t}{y} \tag{8.4}$$

where the subscripts indicate partial derivative.

As in Chapter 1, λ in (8.2) defines the Hicksian rate of technological change and α represents the Harrodian rate. In general, $\lambda = \lambda(k, t)$ and $\alpha = \alpha(v, t)$. These two rates are related to each other as follows. From $y = f(vy, t)$, on differentiating partially with respect to time, we have $h_t = \pi h_t + f_t$. Hence

$$\alpha = \frac{1}{1 - \pi} \lambda \tag{8.5}$$

Most of what follows below is concerned with developing possible economic interpretations of the two rates λ and α. We begin first with the Hicksian rate λ.

Differentiating $Y = F_K K + F_L L$ fully with respect to time yields, after taking (8.2) into account,

$$\lambda = \pi \mathring{F}_K + (1 - \pi) \mathring{F}_L \tag{8.6}$$

Under perfect competition the marginal factor products F_K and F_L are equal to factor (real) prices, denoted r and w respectively. The Hicksian rate of technological change, given by either (8.2) or (8.6), can be expressed in the following dual way:

$$\lambda = \mathring{Y} - \{\pi \mathring{K} + (1 - \pi) \mathring{L}\} = \pi \mathring{r} + (1 - \pi) \mathring{w} \tag{8.7}$$

The first of the two representations, if rewritten in terms of proportional changes in average factor productivities, also has the form

$$\lambda = \pi(\mathring{Y/K}) + (1 - \pi)(\mathring{Y/L}) \tag{8.7a}$$

which, as it is equal to the weighted sum of the partial average factor productivity growth rates, can be interpreted to represent the proportional change in the joint average factor productivity.

The cost of producing Y equals $rK + wL$. Call it C. Given input prices, the proportional change in total cost is therefore equal to $\pi \mathring{K} + (1 - \pi) \mathring{L}$. If we define the average cost as the ratio C/Y of total cost to output, the proportional change in this ratio is equal to $\mathring{C} - \mathring{Y}$, which is seen from (8.7) to be equal to $-\lambda$. It follows that λ is either a proportional reduction (increase) in average cost (joint factor productivity), given input prices, or a proportional increase in the income of inputs or the volume of output if the quantities of inputs are held constant.

Let $p = F_L/F_K$, which is equal to the marginal rate of substitution of the two factors $-dK/dL$, when output is kept constant, as well as to the ratio w/r of the factors' competitive demand prices. Since both w and r are functions of k and t, we have $p = p(k, t)$. If p and k are in a one-to-one relationship, as they are in the neoclassical case, then the competitive ratio K/L to be chosen by firms is a function of the p ratio and time t, where the latter represents technology, i.e.

$$k = k(p, t) \tag{8.8}$$

Define

$$\sigma = \frac{p}{k} \frac{\partial k(p, t)}{\partial p} \qquad B^{HI} = \frac{1}{k} \frac{\partial k(p, t)}{\partial t} \tag{8.9}$$

Hence, by differentiating (8.8) totally with respect to time, we have

$$\mathring{k} = \sigma \mathring{p} + B^{HI} \tag{8.10}$$

Given technology, a 1 per cent change in relative factor prices, according to (8.10), would induce a competitive firm to change K/L by σ per cent. The parameter σ thus represents the capital-to-labour elasticity of substitution; it reflects the ease with which one factor can be substituted for another, as firms select new techniques of making things from the given set of production processes in response to a change in factor prices. The term $\sigma \mathring{p}$ represents the size of the substitution effect (in the Hicksian sense).

Given factor prices, the competitive firm can also change the ratio K/L in response to the arrival of new production processes, and B^{HI} in (8.10) measures such a change in percentage terms. We shall say that technological progress is Hicks neutral, Hicks capital-saving, or Hicks labour-saving depending on whether $B^{HI} = 0$, $B^{HI} < 0$, or $B^{HI} > 0$ respectively. It is important to note that the size of B^{HI} has nothing to do with the magnitude of the rate of technological change, but only indicates the presence of a particular bias in that change. Thus a new production process would be Hicks capital-saving ($B^{HI} < 0$) if its appearance induces the cost-minimizing firm to reduce the capital-to-labour ratio, even though the factor price ratio remains unchanged.

Since $k = K/L$ and $p = F_L/F_K$, $k/p = F_K K/F_L L = \pi/(1 - \pi)$ = profits/wages. Hence, given the p ratio, any Hicks capital-saving innovation reduces not only k but also the capital's (competitive) income share. On differentiating this relation between k and π partially with respect to time, keeping either p or k constant, it follows that

$$\frac{1}{\pi} \frac{\partial \pi(p,\, t)}{\partial t} = (1 - \pi)B^{\mathrm{HI}}$$

or

$$\frac{1}{\pi} \frac{\partial \pi(k,\, t)}{\partial t} = \frac{1}{\sigma}(1 - \pi)B^{\mathrm{HI}}$$

(8.11)

where we make use of the notation given in (8.9).

A production function can also be defined implicitly by specifying the technological rate λ as a function of k and t. The reader can check that the following relation holds:

$$\operatorname{sgn} \frac{\partial \lambda(k,\, t)}{\partial k} = \operatorname{sgn} \frac{\partial k(p,\, t)}{\partial t}$$

(8.12)

This represents the relation between the Hicksian bias of technological change in response to a change in the capital-to-labour ratio. We can thus think of the Hicksian bias in terms of this response, as illustrated in Figure 8.1.

As noted above, Hicks defined the substitution effect and the technological bias effect in terms of changes in the ratio(s) of inputs. Harrod, however, defined these two effects in terms of changes in the ratios of inputs to outputs. The latter definition has been widely adopted by the so-called (English) Cambridge School, in particular by Joan Robinson and Michal Kalecki. The definition has also proved, for some purposes which will be discussed further later, to be more relevant than the Hicksian definition. Suppose, then, that we take v as the relevant input-to-output ratio. Since $r = r(k,\, t)$ and $v = k/f(k,\, t)$, under suitable (e.g. neoclassical) assumptions we can write

$$v = v(r,\, t)$$

(8.13)

In a two-inputs one-output case, there is almost complete symmetry between the Harrodian analysis and the Hicksian analysis, with v and r in the former paralleling k and p in the latter. In particular, similarly to (8.10), we have

$$\mathring{v} = -\sigma \mathring{r} + B^{\mathrm{HA}}$$

(8.14)

where $B^{\mathrm{HA}} = (1/v)\partial v(r,\, t)/\partial t$ represents the size of the Harrodian bias in the technological change and $\sigma \mathring{r}$ is the Harrodian substitution term. We define technological progress as Harrod

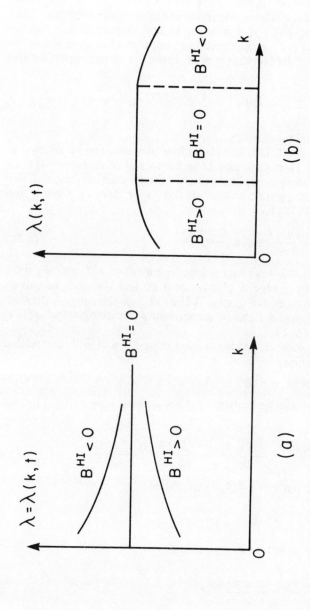

Figure 8.1 The Hicksian bias and the responsiveness of the Hicksian rate of technological progress to a change in the capital-to-labour ratio: (a) the same type of bias for all k; (b) the type of bias may change from one range of k to another.

123

neutral, Harrod capital-saving, or Harrod labour-saving, depending on whether $B^{HA} = 0$, $B^{HA} < 0$, or $B^{HA} > 0$ respectively. Thus, given r, a Harrod capital-saving innovation induces a competitive firm to reduce the (cost-minimizing) capital-to-output ratio. Alternatively, given r, a new production process that is capital-saving would reduce the capital's (competitive) income share, the implication of the fact that $\pi = rv$. From the latter relation we also have

$$\frac{1}{\pi} \frac{\partial \pi(r, t)}{\partial t} = B^{HA} \qquad \frac{1}{\pi} \frac{\partial \pi(v, t)}{\partial t} = \frac{1}{\sigma} B^{HA} \qquad (8.15)$$

which parallel relations (8.11).

Similarly to (8.12), the Harrodian bias can also be defined in terms of the Harrodian rate of technological change $\alpha = \alpha(v, t)$. The latter function implicitly defines the production function, and hence also the profit share function $\pi = \pi(r, t)$. Chilosi and Gomulka (1974) have shown that

$$\text{sgn} \frac{\partial \alpha(v, t)}{\partial v} = \text{sgn} \frac{\partial \pi(r, t)}{\partial t} \qquad (8.16)$$

From (8.16) it follows that α is independent of v if and only if the technological change is Harrod neutral, and that it is negatively related to v when the change is Harrod capital-saving. A diagram similar to Figure 8.1 can be drawn, with v replacing k and $\alpha(v, t)$ replacing $\lambda(k, t)$.

The two measures of innovation bias, B^{HI} and B^{HA}, are related by the equation

$$B^{HA} = (1 - \pi)B^{HI} - (1 - \sigma)\lambda \qquad (8.17)$$

This relation can be derived as follows. From $\pi = \pi\{v(k, t), t\}$ we have

$$\frac{\partial \pi(k, t)}{\partial t} = \frac{\partial \pi(v, t)}{\partial v} \frac{\partial v(k, t)}{\partial t} + \frac{\partial \pi(v, t)}{\partial t}$$

However, from $v = k/f(k, t)$ we obtain

$$\frac{\partial v(k, t)}{\partial t} = -\lambda v$$

and from $p = k(1 - \pi)\pi$ we obtain

$$\frac{\partial \pi(v, t)}{\partial v} = \frac{\pi}{v} \frac{\sigma - 1}{\sigma}$$

Substitution of these two terms and reference to equations (8.11) and (8.15) give (8.17). From (8.17) we note that, for $\sigma = 1$, the direction of bias is the same under both classifications. It should also be noted that, in the absence of technological change, λ, B^{HI}, B^{HA} are all zero.

The percentage change in capital's competitive income share is, not surprisingly, also a sum of two terms, one representing a substitution effect and the other the effect of an innovation bias. The expressions for these two effects depend on their Hicks and Harrod definitions, as is seen explicitly from equation (8.18) giving their joint effect on $\overset{\circ}{\pi}$:

$$\overset{\circ}{\pi} = \frac{1 - \pi}{\sigma} \{(\sigma - 1)\overset{\circ}{k} + B^{HI}\} = \frac{1}{\sigma} \{(\sigma - 1)\overset{\circ}{v} + B^{HA}\} \quad (8.18)$$

From (8.18) it follows that if, with time, the capital-to-output ratio is constant and the capital-to-labour ratio is rising, the substitution effect on π will be zero according to Harrod's definition but, for $\sigma \neq 1$, it will be non-zero according to Hicks's definition.

Under non-constant returns to scale, most of the relations given above do not hold, for the size of the technological bias, whether Hicksian or Harrodian, would then also depend on the scale of output. The measures of bias must in this case be defined as follows:

$$B^{HI} = \frac{1}{k} \frac{\partial k(p, y, t)}{\partial t} \qquad B^{HA} = \frac{1}{v} \frac{\partial v(r, Y, t)}{\partial t} \quad (8.19)$$

The elasticity of capital-to-labour substitution can be defined similarly. The (optimal) factor ratios K/L and K/Y may change, under non-constant returns to scale, in response not only to technological change and factor prices p and r respectively but also to the scale of output.

If returns to scale are unevenly distributed among production factors, then the underlying production function is said to be non-homothetic and a bias in the returns to scale is said to be present. The scale bias S can be defined as follows:

$$S^{HI} = \frac{1}{k} \frac{\partial k(p, Y, t)}{\partial Y} \qquad S^{HA} = \frac{1}{v} \frac{\partial v(r, Y, t)}{\partial Y} \quad (8.20)$$

i.e. S in (8.20) is defined in terms of the relative changes in (optimal) factor ratios that would take place, given prices and technology, in response to a change in the scale of output.

The n-factor case

Suppose that $Y = F(X_1, \ldots, X_n, t)$, where the Xs denote the quantities of inputs and Y is the quantity of a homogeneous output. Between a pair of factor inputs, i and j say, the technological bias in the Hicks sense can be defined as

$$B_{ij}^{\text{HI}} = \left. \frac{\partial \ln(X_i/X_j)}{\partial t} \right|_{p_i/p_j, Y \text{ fixed}} \qquad i, j = 1, \ldots, n \qquad (8.21)$$

where p_i/p_j is the ratio of the factor (demand) prices. As in the two-factor case, we note that this Hicksian measure of bias is a property not of each factor but of each pair of factors. For instance, if the factors are capital, labour, and energy, technological change may be energy-saving relative to capital (given p_E/p_K, the ratio E/K would be declining with time) but energy-using relative to labour (given p_E/p_K, the ratio E/L would be increasing with time). Thus we would have $n - 1$ measures of bias, or bias functions, for so many pairs, all involving the *same* factor.

In empirical estimation, this multitude of measures may be a disadvantage. One possible alternative is to define bias for each factor in terms of relative changes in the factor's share, with the factor prices and the scale of output being held constant (Binswanger 1974b; in this study of technological biases in American agriculture, Binswanger assumes a constant-returns-to-scale function that is separable in terms of output and prices, in which case the factor shares are scale independent).

We would also have only one measure of technological bias for each factor under a definition of the Harrod type:

$$B_i^{\text{HA}} = \frac{\partial \ln(X_i/Y)}{\partial t} \qquad \text{all } p_i \text{ and } Y \text{ fixed} \qquad (8.22)$$

In the case of production processes involving three or more inputs, a number of alternative definitions for the elasticity of substitution have been suggested. These definitions differ in the elasticity's economic interpretation, even though they collapse into the same result in the two-input case. The direct or short-term elasticity of substitution between a pair of factor inputs i and j is defined similarly to (8.9), as a proportional response in X_i/X_j to a 1 per cent change in the factor's price ratio p_i/p_j, holding fixed the supply of all other factors, their prices, and the quantity of output(s). This is the definition which we use in this book. (For a full review of different definitions, see McFadden (1978).)

The Diamond–McFadden–Rodriguez non-identification theorem

The key difficulty in any attempt to estimate empirically the rate and bias of technological change at the level of firms, industries, and whole economies lies in the fact that actual production possibility frontiers, or actual production functions, are not observable. Even if these functions were non-stochastic and of the neoclassical type, and if the firms were always profit maximizers, we would typically know only a single point of any such production function at any given time. Only if more points were known at each time could the elasticity(ies) of factor substitution be estimated independently of the technological bias. As it is, the economist is left to draw inferences about σ and bias from the time series data only. By equation (8.10),

$$B^{HI} = \mathring{k} - \sigma\mathring{p}$$

in which $\mathring{k}$ and $\mathring{p}$ are known. If p is constant, the value of σ is immaterial and the bias can be measured by observing changes in k alone. In the other limiting case, when technological progress is absent, the bias is zero and, in turn, the elasticity is measurable. However, since p is typically changing and there is no a priori information about the bias, the elasticity and bias cannot be identified simultaneously.

Diamond, McFadden, and Rodriguez phrased this fact in the following way:

> Given positive regular observed $y^*(t)$, $k^*(t)$, $p^*(t)$, $0 \leqslant t \leqslant t_1$, for output per worker, capital per worker, and the ratio of marginal products, respectively, generated by a neoclassical production function $Y = f(k, t)$ which exhibits positive technical progress on the observed path, and given any positive continuously differentiable function $\sigma^*(t)$ on $0 < t < t_1$, there exists a neoclassical production function $f(\tilde{k}, t)$ which generates the observed series and has an elasticity of substitution $\sigma^*(t)$ along the path $(k(t), t)$. (McFadden 1978: 133, Theorem 1; the original statement of the theorem is due to Diamond *et al.* 1965)

Their proof of the statement is quite involved, but its content can be grasped intuitively from the following representation of production functions (McFadden 1978: Chapter IV.1). From the definition of the ratio of marginal products we have $p(k, t) = -k + f/f_k$, and hence $f_k/f = \{p(k, t) + k\}^{-1}$. Upon integrating both sides of the latter equation with respect to the capital-to-labour ratio, from the actually observed $k^*(t)$ to any arbitrary k, we obtain

$$\ln\{f(k, t)\} = \ln\{y^*(t)\} + \int_{k^*(t)}^{k} \{p(z, t) + z\}^{-1} \, dz \qquad (8.23)$$

where $y^*(t) = f\{k^*(t), t\}$ and is also actually observed. Similarly, on integrating the definitional equation (8.8), the price ratio p can be expressed in terms of σ as follows:

$$\ln\{p(k, t)\} = \ln\{p^*(t)\} + \int_{k^*(t)}^{k} \frac{\sigma(z, t)^{-1} \, dz}{z} \qquad (8.24)$$

where $p^*(t) = p\{k^*(t), t\}$ and is observed. The unobserved production function $f(k, t)$ is seen to be defined fully by the observed time series data $y^*(t)$, $k^*(t)$, and $p^*(t)$, and the unobserved elasticity function $\sigma(k, t)$. Thus, by selecting different $\sigma(k, t)$ functions, we can generate different production functions, all of which would be consistent with both the given data and the condition of non-retrogressive technological change on the observed growth path. (In this section, care must of course be taken to ensure that the production functions generated in this way exhibit suitable neoclassical properties.)

In order to obtain a unique production function consistent with the observations, in addition to displaying the neoclassical assumptions, some further restrictions must be imposed. The property of factor augmentation has been shown by Diamond and co-authors to be insufficient, except in some cases when only one factor is being augmented. However, Sato (1970) shows that a still further restriction, in the form of selecting σ to be either a constant or dependent on the capital's share only, will permit identification. Presumably there are many other restricting assumptions which would be equally effective.

Technological bias and the question of balanced growth

A process of economic growth can be said to be balanced (or proportional or steady) if certain key ratios remain constant over time. Depending on the choice of the ratios, some definitions are more restrictive than others. Usually, however, it is required (i) that the quantities of all outputs and producible inputs expand at a percentage rate which is common and constant, (ii) that the quantities of all the non-producible inputs, in particular labour inputs, also expand at constant, but possibly input-specific, percentage rates, and (iii) that all the input–output ratios remain constant in value terms. The time period to be considered may also vary. For practical reasons a finite period should suffice, but it may include all past and future, or all future only.

The very fact of technological innovation, by which some inputs disappear and new ones come into play, means that real economies do not meet these requirements; the growth paths they follow are always unbalanced. However, it has been observed that at high levels of aggregation, such as large sectors and whole economies, the capital-to-output ratios tend to be relatively stable over periods of considerable length in both constant and current prices, while the labour-to-output ratios tend to decline consistently at approximately constant growth rates. This characteristic of the real world indicates that the prevailing technological changes are of a certain category. What category?

Let us consider the aggregate (one-sector) economy case first. Suppose that $Y = F(K, L, t)$, $\dot{K} = sY$, where s is a given constant, $0 < s < 1$, and $\mathring{L} = n$, which is another constant. Hence, by equation (8.2), $\mathring{y} = q\mathring{v} + \alpha$, and $s = \dot{K}/Y = v(\mathring{y} + n) + \dot{v}$.

Under balanced growth both $\mathring{y}$ and v must be constant. Therefore, given $v = v^*$, the rate of technological change α must be constant, but by (8.16) we note that to meet these conditions the bias of technological change need not be Harrod-neutral. What is required of the function $\alpha(v, t)$ is only that $\alpha(v^*, t)$ is constant for all t over any (finite) period considered, where $v^* = s/\{\alpha(v^*, t) + n\}$, or, if balanced growth is to take place for a range of values of the (constant) investment ratio, that $\alpha(v, t)$ is time independent for the corresponding range of values of the capital-to-output ratio. In the latter case, it is necessary that $\alpha = \alpha(v)$ for all v from the range in question, with α independent of v (the Harrod-neutral technological change), as a special case.

However, in the presence of the Harrod non-neutral bias, the competitive 'rate of profit' is not constant and the competitive wage rate is not changing at a constant rate. If (iii) above is added to (i) and (ii) as a requirement of balanced growth, then such growth process would be permitted only by technological progress such that the rate α is invariant with respect to v, i.e. the innovations are purely labour saving. In fact, the actual wage share has been observed to be approximately constant in some sectors, economies, and periods of time. When it is, and if all wages are equal, at least approximately, to labour's marginal products, the above technological condition gives grounds for believing that the underlying technological change is, in fact, predominantly labour saving.

We shall often return to this important point later in the book, particularly in the next chapter. (For a complete discussion of the technological requirements for balanced growth in the context of a one-sector (aggregate) economy, see Whitaker (1970) and Chilosi and Gomulka (1974).)

The n-sector case

Let us now consider an economy consisting of n sectors, each producing goods of one type, of which $n - 1$ are producing capital goods and one, the nth sector, is producing single-consumption goods. Each sector is characterized by a constant-returns-to-scale production function with technological change of the factor-augmenting type. Thus, in a self-explanatory notation, we can write

$$Y_j = F^j(A_{ij}K_{ij}, \ldots, A_{n-1,j}K_{n-1,j}; B_j N_j) \qquad j = 1, \ldots, n$$

(8.25)

where N_j is the quantity of the single non-reproducible (labour) input in sector j. The feasibility of this production structure's being capable of expanding in a balanced way has been studied by Charles Kennedy (1973). He has defined balanced growth as a state of the economy when (i) all inputs, outputs, prices, and augmentation factors change at constant rates, (ii) the output composition of the economy and the input composition in each sector remain unchanged when measured in value terms, and (iii) the marginal product of capital, or the rate of interest, is constant and the same in all sectors for all capital goods. It can be seen that this definition realistically permits a change in the composition of output in real terms.

To ascertain the type of technological change that would permit the balanced growth as defined above, we note first that all sectoral labour inputs must grow at a common rate n, wages must grow at another common rate $\mathring{w}$, and all K_{ij}/Y_i must remain constant. With these properties in mind, we obtain from (i) and (iii)

$$\mathring{Y}_i + \mathring{A}_{ij} = n + \mathring{B}_j = \mathring{Y}_j \tag{8.26}$$

and from (ii)

$$\mathring{Y}_i + \mathring{p}_i + n + \mathring{w} = \mathring{Y}_j + \mathring{p}_j \tag{8.27}$$

where p_i is the price of the ith good. Since $\mathring{Y}_j = \mathring{B}_j + n$, the sectoral output growth rates are seen to be decided, given n, by the rates at which the productivities of the non-producible input, labour in this case, are rising in the different sectors. We also have that $\mathring{p}_i - \mathring{p}_j = \mathring{B}_j - \mathring{B}_i$. Suppose that the price p_n of the consumer good is unity, so that w is also the real wage rate. Hence $\mathring{p}_i = \mathring{B}_n - \mathring{B}_i$, or the relative price changes are equal to the corresponding changes in the relative labour-to-output ratios. However, the technological change need not be purely labour augmenting. From (8.26) it follows that $\mathring{A}_{ij} = \mathring{B}_j - \mathring{B}_i = \mathring{p}_i - \mathring{p}_j$. Hence $\mathring{A}_{ij} - \mathring{A}_{ik} = \mathring{B}_j - \mathring{B}_k = \mathring{p}_k - \mathring{p}_j$. Thus, if a sector j of Kennedy's steady state economy is relatively progressive compared with a sector k, so that the price ratio p_j/p_k is falling, then the productivity of the labour input and the (partial) productivities of the capital input, measured on a balanced growth path by $\mathring{B}_j$ and $\mathring{A}_{ij}$ respectively, would all have to be increasing relatively faster in that sector than in sector k.

Technological change must be of a very particular type to meet the Kennedy requirements (8.26) and (8.27). It is legitimate and proper to insist, as Rymes (1973) does, that this type of change can still be called Harrod-neutral. The trouble is that 'Harrod-neutrality' is usually identified with purely labour-augmenting technological progress, especially for the case of a single-good economy. For this reason we shall follow Kennedy (1973) himself who suggests calling it neutral in the Harrod–Kennedy sense for the general case of many capital goods.

As we have noted above, $\mathring{A}_{ij} = \mathring{B}_j - \mathring{B}_i$. Since $\mathring{B}_i > 0$ for all i, it follows that if each capital good is needed, directly or indirectly, in the production of every good, and if technological progress is taking place in all sectors, then

$$\mathring{B}_j > \mathring{A}_{ij} \qquad \text{for all } i, j \qquad (8.28)$$

The Harrod–Kennedy neutrality therefore has the property that innovations are predominantly labour-saving (in general, saving predominantly non-producible resources).

This important property can be given still additional emphasis for an economy that moves along an optimal (balanced) growth path in which labour (and/or other non-producible inputs) cannot be substituted easily for capital inputs (this point is discussed further below).

We recall that, in the one-sector case, the balanced-growth capital-to-output rate is constant when measured in volume and value terms. In our n-sector case, the sectoral capital-to-output ratios are also constant on a balanced-growth path but, in principle, only when measured in value terms. Since the sectoral value output composition does not change, the same is true of the aggregate capital-to-output ratio. The actually observed capital-to-output ratios, both sectoral and aggregate, apparently tend to be quite stable compared with the consistently declining labour-to-output ratios, but this observed relative stability appears to be present irrespective of whether the ratios are measured in constant

131

or current prices. It is therefore of interest to ask for the circumstances under which the system (8.26) and (8.27) would also display this property. The relevant conditions are implied by the requirement that all ratios K_{ij}/Y_j must be constant. Hence $\mathring{Y}_i = \mathring{Y}_j$ and, by (8.26), all $\mathring{A}_{ij} = 0$, while all $\mathring{B}_i$ would have to be the same. Thus this is the rather restrictive case of technological innovations being purely labour-saving in the Harrod sense, and innovation rates being the same in all sectors. However, if the rate $\mathring{B}_i$ varies considerably among sectors, as it apparently does, the long-term stability of the sectoral capital-to-output ratios is something of a paradox. We shall come back to this issue in the last section of this chapter.

The measurement of productivity change and its contribution to growth: a disaggregated analysis and aggregation rules

The mere number of newly registered patents, or new inventions which have been identified in some way, is widely thought to be deficient as a direct (aggregate) measure of inventive activity, although it does provide a rough indication of its size and is in fact often used for that purpose. It is a deficient measure since, in addition to the time-varying composition of innovations in terms of their economic significance, such measures ignore what is essential, i.e. the rate of the actual application of the innovations and the extent of their diffusion. Indirect measures must therefore be used instead, whereby the size of the flow of innovation would be gauged from their economic effect. However, with any indirect measures we immediately encounter three major problems: (i) defining the economic effects; (ii) separating out the innovation-induced effect from all other economic effects; (iii) the length of time to be considered, and hence the extent to which the economic effects that are caused only indirectly by innovations should also be taken into account. We have touched upon these problems already in Chapter 1.

Bearing in mind the ultimate purpose of economic activity, possibly the best measure of economic effects of innovation processes would be the totality of the corresponding change in individual utilities. Such a measure would also have the advantage that it could handle the effects of product innovation and of the innovation-induced environmental changes. However, for obvious practical reasons, the method which has been used widely, and which will also be adopted below, is one which measures the economic effects in terms of 'real output' or 'real consumption'.

Also for practical reasons, we shall make no attempt to separate out the contribution of innovations from that of the other qualitative changes, such as upgrading human skills. Clearly, in reality, there is typically a complex interplay among the different kinds of qualitative changes, especially over longer periods of time. Our proposed measures of the contribution of technological innovation to growth are therefore all-inclusive, capturing also the contribution of these other qualitative changes. In the long run, it is in any case probable that neither technological innovation nor job training is likely to occur or be effective as a source of economic growth in the absence of the other type of qualitative change.

The reproducibility problem

In Chapter 1 we noted that any estimate of the contribution of technological change to economic growth over a period of time must also take account of the contribution of those quantities of inputs which were produced in the economy as a result of technological change during that period. This contribution is what we have called the indirect or induced effect of technological change. Only if all inputs were non-producible or primary could technological change increase the flow of outputs exclusively through reducing the use of some or all of such inputs per unit of output, which is the direct effect.

However, most inputs, in particular intermediate and capital inputs, are in fact producible. Even land can be reclaimed and population size can respond to changes in consumption.

In a multisectoral context, the distinction can also be made between the direct effect on the output of a particular sector of the technological change that, at any given moment of time, originates in that sector, and the total output effect in the same sector of the technological change that takes place in the whole economy. The formulae which we shall develop for the sectoral rates of technological change take into account inter-industrial flows of producible inputs, since these inputs are the agents which transmit the output impacts of any innovation from its adopters to the users of the adopters' outputs.

The Hicksian, the Harrodian, and the Fisherian rates of technological change

Let us consider first the case of an (aggregate) one-sector economy in which the difference between the Hicksian rate λ and the

Harrodian rate α has already been discussed in detail. This difference is also shown in Figure 8.2.

Suppose that initially the economy is at a point A. If k^* is the equilibrium K/L in a constant-returns-to-scale economy, with no change in technology and a constant savings ratio, the economy would remain at A indefinitely. However, when technological changes take place, the production function $f(k, t)$ shifts upwards and the economy moves along a growth path that originates at A. The difference between the output per man y in the latter case, and the equilibrium level y^* in the former case, may therefore be attributable to these technological changes. Unfortunately a complication immediately arises, for the size of this difference also depends on the growth rate of employment and the fraction of output that is being added to the capital stock along the economy's

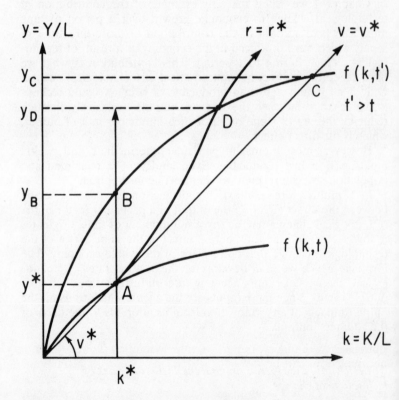

Figure 8.2 The Hicksian path AB, the Harrodian path AC, and the Fisherian path AD.

growth path, and these two parameters could be partly influenced, directly and indirectly, by a number of factors, some of them non-economic.

Suppose that we regard the observed growth rate of employment as a given parameter and specify, for the innovative economy, a 'proper' savings ratio. Three choices of the ratio, out of the many possible, have attracted the particular interest of economists.

The choices are related to the fact that in any innovation-free economy the capital-to-labour ratio k^*, the capital-to-output ratio v^*, and the interest rate r^* are all constant. A 'proper' savings ratio is defined by the requirement that one of these three steady state characteristics also remains constant in the innovative economy. This requirement gives rise to three different savings ratios and therefore to three alternative growth paths, marked in Figure 8.2 by AB, AD, and AC. Consequently, we have three different measures of the output effect of technological change: the Hicksian λ, the Harrodian α, and the Fisherian ψ respectively.

These three alternative growth paths need not be socially optimal nor balanced. Take, for instance, the AC path along which the capital-to-output ratio is constant. The rate of growth of the economy's output along this path is a sum of two components, n and α, both of which may be changing over time so that their sum need not be constant. Since $(\partial \alpha / \partial v)_{v=v^*}$ may not be zero, the technological change along the path need not be Harrod-neutral either.

How can we justify the choice of one characteristic in preference to the other two? The Harrodian choice would be well justified if and when the actual technological change is close to being Harrod-neutral, since the actual capital-to-output ratio should then be approximately constant and hence the actual growth path is close to the Harrodian path. The case for requesting the constancy of the interest rate is made on the grounds that, in the absence of technological change, an optimal growth path (the one which maximizes the total discounted stream of future utilities) tends to a stationary state in which the marginal rate of capital, or the rate of interest, is equal to the time preference rate which is assumed to be constant (Hulten 1975). However, this result does not hold for all cases with technological change present. The optimal growth theory, in particular the Modified Golden Rule of accumulation, can be invoked to note that the optimal marginal product of capital tends to a constant only when the technological progress is Harrod neutral. However, in this particular case, the Harrodian α and the Fisherian ψ are in fact the same. In the general case, there is no good justification for requiring r or v to

135

be constant, and still less for requiring k to be constant.

We end this section by noting the relations between α, ψ, and λ. The relation between α and ψ can be obtained as follows. By definition, $\psi = \mathring{f}_{r=r^*}$. Since $\mathring{f} = \pi \mathring{k} + \lambda$ and $\pi = rv$, we have, by substituting for $\mathring{\pi}$ and $\mathring{k}$ in equation (8.18) and rearranging,

$$\psi = \alpha + \frac{\pi}{1 - \pi} B^{\mathrm{HA}} \tag{8.29}$$

which confirms that under Harrod neutrality $\psi = \alpha$.

In Chapter 1 we established the relation between the Hicksian rate λ and the Harrodian α by reference to the 'growth propagation effect'. Another way of obtaining the same relation has been suggested by Rymes (1971, 1973). It rests on the concept of 'real capital', which is measured in terms of the non-producible resources – labour hours in our present case – that would be required to reproduce the (commodity) capital stock at any particular moment, if the economy has travelled along a growth path on which the ratio of capital stock to output is constant, say v^*. In order to obtain an index for real capital in a one-sector constant-returns-to-scale economy, the (commodity) capital stock K must be deflated by the labour productivity index y^* which is equal to $f(v^*, t)$. It follows that the growth rate of real capital in the economy in question is equal to $\mathring{K} - \alpha$, where α is the growth rate of $f(v^*, t)$ and $\mathring{K}$ is the observed growth rate of the capital stock. We have, on the v^* path, that $\mathring{Y}^* = \mathring{K}^* = n + \alpha$, and, on the actual path, that $\mathring{Y} = \pi(\mathring{K} - \alpha) + (1 - \pi)n + \alpha$. However, $\mathring{Y} - \pi\mathring{K} - (1 - \pi)n = \lambda$. Hence $\alpha = \lambda/(1 - \pi)$.

In the case of many non-producible (or primary) inputs and many sectors, each producing one type of output, the Rymes method of measuring the sectoral and aggregate rates of technological change is also based on the principle that they 'reflect the changing efficiency by which primary inputs are being transformed into outputs' (Rymes 1973: 97). Thus, suppose that

$$Q_j = F^j(X_{1j}, \ldots, X_{rj}; K_{r+1,j1}, \ldots, K_{nk}; N_{1j}, \ldots, N_{mj}; t)$$

$$j = 1, \ldots, n \tag{8.30}$$

where X_{ij} represents an intermediate input of the ith good, K_{sj} is a capital input of the sth good, and N_{kj} stands for the kth primary input. Let $q_{kj} = Q_j/N_{kj}$. Then the Harrodian productivity change for sector j is

$$\alpha_j = \mathring{Q}_j - \sum_{k=1}^{m} \omega_{kj}\mathring{N}_k \bigg|_{\text{HK growth path}} = \sum_{k=1}^{m} \omega_{kj}(Q_j/N_k)\bigg|_{\text{HK}} \tag{8.31}$$

where $\omega_{kj} = \omega_k N_{kj} / \Sigma \; \omega_k N_{kj}$. Thus α_j is the growth rate of the joint average productivity of the primary inputs in sector j and is equal to the weighted sum of the average productivity growth rates of the individual inputs on the Harrod–Kennedy growth path, on which X_{ij}/Q_i, K_{sj}/Q_s, and N_{kj}/N remain constant for all i, s, and k, with factor shares as weights. (However, we stress again that the path need not be one of balanced growth, since the growth rates of outputs and inputs need not be constant. The reader may note, however, that the Kennedy balanced-growth path has the properties of the Harrod–Kennedy path. Hence, in the Kennedy model $\alpha_j = \beta_j$, the result which is implied by equation (8.26).)

The relation between the Hicksian rates $\lambda_1, \ldots, \lambda_n$ and the Harrodian rates $\alpha_1, \ldots, \alpha_n$, for the economy described by (8.30), can be found as follows. From the value output–cost identity

$$p_j Q_j = \sum_{i=1}^{r} p_i X_{ij} + \sum_{s=r+1}^{n} p_s r K_{sj} + \sum_{k=1}^{m} \omega_k N_{kj} \tag{8.32}$$

we find

$$\mathring{p}_j + \mathring{Q}_j = \sum_i a_{ij}(\mathring{p}_i + \mathring{X}_{ij}) + \sum_s b_{sj}(\mathring{p}_s + \mathring{r} + \mathring{K}_{sj}) + \sum_{k=1} v_{kj}(\mathring{\omega}_k + \mathring{N}_{kj})$$

where $a_{ij} = p_i X_{ij}/p_j Q_j$, $b_{sj} = r p_s K_{sj}/p_j Q_j$, and $v_{kj} = \omega_k N_{kj}/p_j Q_j$. By definition,

$$\lambda_j = \mathring{Q}_j - \sum_i a_{ij}\mathring{X}_{ij} - \sum_s b_{sj}\mathring{K}_{sj} - \sum_k v_{kj}\mathring{N}_{kj} \tag{8.33}$$

or, in terms of prices,

$$\lambda_j = \mathring{p}_j - \sum_i a_{ij}\mathring{p}_i - \sum_s b_{sj}(\mathring{p}_s + \mathring{r}) + \sum_k v_{kj}\mathring{\omega}_k \tag{8.33a}$$

We now deflate all producible inputs by the joint productivities of the non-producible inputs in particular sectors, so that the former are valued in terms of the latter. Hence

$$\alpha_j = \mathring{Q}_j - \sum_i a_{ij}(\mathring{X}_{ij} - \alpha_i) - \sum_s b_{sj}(\mathring{K}_{sj} - \alpha_s) - \sum_k v_{kj}\mathring{N}_{kj} \tag{8.34}$$

or, in terms of prices,

$$\alpha_j = -\mathring{p}_j - \sum_i a_{ij}(\mathring{p}_i - \alpha_i) - \sum_s b_{sj}(\mathring{p}_s + \mathring{r} + \alpha_s) + \sum_k v_{kj}\mathring{\omega}_k \tag{8.34a}$$

137

In terms of both quantities and prices we have, then,

$$\alpha_j = \lambda_j + \Sigma\, a_{ij}\alpha_i + \Sigma\, b_{sj}\alpha_s \qquad (8.35)$$

This is the required relation. The Leontief input–output structural matrices $A = \{A_{ij}\}$ and $B = \{b_{ij}\}$, both in value terms, are usually known, and so (8.35) can be used for the purpose of solving explicitly for α. Let

$$A^G = \begin{Bmatrix} A \\ B \end{Bmatrix}$$

an $n \times n$ technology matrix. From (8.35) we have that $\alpha = \lambda + A^G\alpha$. Hence

$$\alpha = (I - A^G)^{-1}\lambda \qquad (8.35a)$$

It is tempting to interpret λ_j in (8.35) as a measure of the technological change which originates in sector j only, in contrast with α_j which also reflects the changes that originate in the input-supplying industries. This interpretation would in fact be correct if product innovation were ruled out and λ_j measured only the size of the cost-reducing process innovation taking place in industry j. However, in real situations λ_j must also be taken to reflect the changes in the qualities of the inputs supplied by the other industries, in which case the structural interdependence of the economy immediately comes into play. The latter is thus a common factor underlying both measures of innovation λ_j and α_j. The difference is elsewhere. The rate α_j captures, in addition to λ_j, the multiplier effects of all the rates $\lambda_1, \lambda_2, \ldots, \lambda_n$ on the output of industry j as direct output effects of the λ give rise, over a period of time, to increased flows of producible inputs and these, in turn, result in additional flows of outputs, and so forth. In view of the definition (8.31), α_j measures not just the rate of growth in the output of industry j given all inputs, as λ_j does, but the rate of growth of the output of sector j, $j = 1, \ldots, n$, which the economy is capable of sustaining over a long period of time, given the supply of only the non-producible inputs.

The gross versus the net innovation rate, and the transfer effect

In any empirical investigation of intersectoral flows of technological change, it is interesting to know what difference it would make to the values of λ and α if sectoral gross outputs are replaced by respective net outputs, i.e. values added. We shall adopt the notation λ^G, α^G and λ^N, α^N for the innovation rates in the two cases. The rate λ_j^G is given by (8.33), while the rate λ_j^N, being defined as a weighted average of the rates of growth of net factor productivities, is

$$\lambda_j^N = \mathring{Y}_j - \left(\sum_s b_{sj} \mathring{K}_{sj} + \sum v_{kj} \mathring{N}_{kj} \right) \frac{Q_j}{Y_j} \qquad (8.36)$$

The relation between λ^G and λ^N can be obtained if we note that the last two terms in (8.32) represent value added in sector j; let it be denoted by $p_j Y_j$. Equation (8.32) can therefore be rewritten in the form $p_j Q_j = \sum p_i X_{ij} + p_j Y_j$. On differentiating this equation totally with respect to time, we obtain

$$\mathring{Y}_j = \frac{Q_j}{Y_j} \left\{ \mathring{Q}_j - \sum_i a_{ij} \mathring{X}_{ij} - \sum_i a_{ij}(\mathring{p}_i - \mathring{p}_j) \right\}$$

Substituting this expression into (8.36) and comparing the result with (8.33) gives us the required relation

$$\lambda_j^N = \frac{Q_j}{Y_j} \{ \lambda_j^G + \sum a_{ij}(\mathring{p}_j - \mathring{p}_i) \} \qquad (8.37)$$

It is significant that price changes enter into this formula. Real value added in industry j is increasing when the relative prices of the intermediate inputs it uses decline. The second term in (8.37), $\sum_i (Q_j/Y_j) a_{ij}(\mathring{p}_j - \mathring{p}_i)$, is a weighted average of the proportional rates of decline of the intermediate input prices. Hence the term represents the proportional rate of decline of the joint intermediate input per unit of value added. The term can also be said to capture a 'net transfer effect' (Azam 1980: 39). When the term is positive, it is as if industry j receives a transfer of value added from the rest of the economy, the value of which may or may not be due to technological change. The latter feature is potentially a serious flaw of the λ^N rate.

Now we turn to the relation between α^G and α^N. We have already found the relation between α^G and λ^G; it is given by (8.34). Applying again the method proposed by Rymes, we have that

$$\alpha_j^N = \mathring{Y}_j - \frac{\{ \sum b_{sj}(\mathring{K}_{sj} - \alpha_s^N) + \sum v_{kj} \mathring{N}_{kj} \} Q_j}{Y_j}$$
$$= \lambda_j^N + \sum b_{sj}^N \alpha_s^N$$

139

Macroeconomics of innovation

Hence

$$\alpha^N = (I - A^N)^{-1} \lambda^N \qquad (8.38)$$

where $\alpha^N = \{\alpha_j^N\}$ and $\lambda^N = \{\lambda_j^N\}$, $j = 1, \ldots, m$, are column vectors and

$$A^N = \begin{Bmatrix} 0 \\ B^N \end{Bmatrix}$$

is an $n \times n$ matrix where $B^N = \{rp_i K_{ij}/p_j Y_j\}$.

In (8.38) the net Hicksian rates can be converted, by using (8.37), into gross Hicksian rates, and the latter, on applying (8.35a), can be expressed in terms of gross Harrodian rates. Little gain is obtained by pursuing the matter further except to note that, in contrast with λ_j^N in (8.37), each α_j^N is potentially dependent on all α_i^G, $i = 1, \ldots, n$, and that, similarly to λ_j^N in (8.37), the net Harrodian rates of innovation also stand to be affected by 'net transfer effects'.

The aggregation rules for sectoral productivity rates

If sectoral rates of innovation are unknown, then the question is which of the weights should be used for obtaining an average or aggregate rate. We shall list below four major aggregation rules, and pause briefly to consider how they can be interpreted.

There are two alternative rules for the Hicksian sectoral rates which are due to Domar (1961a) and Hulten (1978) respectively:

$$\lambda_A^N = \sum_{j=1}^{n} \frac{Q_j}{Y_j} v_j \lambda_j^G \qquad (8.39)$$

$$\lambda_A^N = \sum_{j=1}^{n} \frac{F_j}{Y_j} v_j \lambda_j^F = \sum_{j=1}^{n} v_j^F \lambda_j^F \qquad (8.40)$$

where $v_j = (p_j Y_j)/\Sigma_j p_j Y_j$ is the weight of the jth sector in terms of its net output (or value added) and F_j is the final demand output of the jth sector (see Hulten (1978) for more details). The rate λ_j^F in (8.40) is what Hulten calls the 'effective rate of productivity change'. It is defined as an innovation-caused change in final demand rather than in gross output, given the economy's total supplies of all non-intermediate inputs.

For the case of the Harrodian (gross) rates we have the following aggregation rules due to Rymes (1973) and Azam (1980) respectively:

$$\alpha_A = \sum_{j=1}^{n} \gamma_j \alpha_j \qquad (8.41)$$

$$\alpha_A = \sum_{j=1}^{n} \omega_j \alpha_j \qquad (8.42)$$

where $\gamma_j = p_j Q_j / \Sigma\ p_j Q_j$ represents the weight of the jth sector in terms of its gross output, and $\omega_j = (\Sigma_k\ w_k N_{kj}) / \Sigma_k\ w_k N_k$ represents the weight of the jth sector in terms of the non-producible inputs it employs or uses (see Rymes (1973) and Azam (1980); Azam discusses only the case of a single non-producible input).

The weights add up to more than unity in (8.39) and to unity in the three other formulae. The exception in (8.39) should be intuitively clear if we note that λ_j^G is an output effect of innovation as a fraction of gross output. From (8.37) it follows that, if price effects are ignored, this fraction needs to be multiplied by Q_j/Y_j to express the output effect in terms of net output. We might suspect that the weights would add up to unity if the gross rates in (8.39) were replaced by the net rates with the aid of (8.37). This is indeed the case, but only if a rather subtle requirement is met. For we have

$$\lambda_A^N = \sum_j v_j \lambda_j^N + \sum_j (v_j - v_j^F) \mathring{p}_j \qquad (8.43)$$

where the second term represents the aggregate net transfer effect. This term is seen to be the difference between two proportional price changes for the aggregate net output, one based on value-added weights and the other on final demand weights. Thus the requirement is that the two measures of aggregate price change are the same.

Since the total final demand equals the total value added, we have that $\lambda_A^N = \lambda_A^F$. In view of this equality, rule (8.40) is not surprising; we might have expected the aggregate 'effective rate' to be a weighted sum of the sectoral effective rates, with the sectoral proportional final demands as weights. None the less, the rule is by no means trivial. It holds true irrespective of the bias of technological change, but it requires the assumption that sectoral demand functions for outputs are invariant with respect to the distribution of the fixed total supplies of the primary resources between consumers, which seems a rather strong requirement.

Rule (8.42) allows us to apply our interpretation of the α rate to the economy as a whole, namely that it is the proportional productivity change of all the non-producible inputs if the economy moves along a Harrod–Kennedy growth path.

Optimal bias and Harrod neutrality: innovation possibility frontier versus localized search

A full survey of major stylized facts concerning long-term economic growth, factor substitutability, and innovation rates and biases is given in the next chapter. However, it has been mentioned already that sectoral and economy-wide capital-to-output ratios appear to be fairly stable in both current and constant prices. We have also noted that, in the case of a competitive economy with both returns to scale and distribution shares (approximately) constant, this stability of the capital-to-output ratio would require technological change to be (predominantly) labour saving. How can we account for this bias? In this section we shall present two different answers to this question based on two alternative theories. An argument will also be given why one of the two theories, to be called a localized search theory, might be preferred. A simple model of a single all-purpose good economy is considered first. The implications of the two theories for the case of an *n*-sector economy of the Kennedy type are taken up next.

The Kennedy innovation possibility frontier and Harrod neutrality

Just as Pareto-efficient combinations of conventional inputs and outputs are distinguished to form a production possibility frontier, so also Pareto-efficient (potential) input savings inventions can be distinguished to form an invention frontier. In so far as potential inventions offer us possible alternative innovative choices, the invention frontier also represents the innovation possibility frontier (IPF).

The inventions which are unpredictable in terms of bias form the frontier $E'C'$ in figure 8.3. However, the prevailing prices can be thought of as inducing inventors to direct their intellectual energies towards seeking improvements of a particular kind, namely those which, given the R&D resources, bring the maximum reduction in the cost of supplying the conventional output. This price-induced choice is represented in Figure 8.3 by point D.

The significance of this choice is easiest to grasp if, following Kennedy (1964), Samuelson (1965), and von Weizsäcker (1966), we restrict technological innovation to the class of the 'factor-augmenting' type. In this case we have that

$$Y = F(AK, BL) \tag{8.44}$$

where $A = A(t)$ and $B = B(t)$. Therefore the IPF idea can be modelled by assuming that

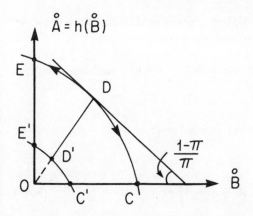

Figure 8.3   Point D represents the (short-term) optimal innovation bias. The bias remains at D if $\sigma = 1$, but it travels towards C if $\sigma < 1$ and towards E if $\sigma > 1$.

$$\mathring{A} = h(\mathring{B}) \tag{8.45}$$

where h has the curvature shown in Figure 8.3 (i.e. h is declining with $\mathring{B}$ and is concave). Given factor prices, the percentage cost reduction is measured by the Hicksian innovation rate λ which in this case is given by

$$\lambda = \pi\mathring{A} + (1 - \pi)\mathring{B} = \pi h(\mathring{B}) + (1 - \pi)\mathring{B}$$

Hence $\mathring{B}$ maximizing λ is given by the condition

$$h' = -\frac{1 - \pi}{\pi} \tag{8.46}$$

where $(1 - \pi)/\pi$ is the ratio of labour to capital costs. Thus while the choice of the ratio K/L itself is given by the price ratio of the factors, the choice of the direction of technological change $\mathring{A}/\mathring{B}$ is determined, in this model, by the cost ratio of the factors. Shifts in the production function (8.44) over time are no longer exogenously given but are driven by such cost-dependent choices. These shifts would in turn influence input prices and therefore the choice of inputs. The economy would thus evolve along a growth path on which the innovation bias may be changing, but in what direction and to what end?

To answer this question, we must close the model by specifying the supply functions for capital and labour. It is convenient to assume that $\mathring{L} = n$ and $\mathring{K}/Y = s$, where both n and s are constant. On applying (8.18) we obtain

$$\mathring{\pi} = (1 - \pi) \frac{\sigma - 1}{\sigma} (\mathring{K} + \mathring{A} - n - \mathring{B})$$

However, $\mathring{K} = s/v$, and

$$\mathring{v} = \mathring{K} - \mathring{Y} = (1 - \pi) \frac{s}{v} - \pi\mathring{A} - (1 - \pi)(n + \mathring{B})$$

or

$$\mathring{v} = (1 - \pi)\left(\frac{s}{v} + \mathring{A} - n - \mathring{B}\right) - \mathring{A} \tag{8.47}$$

It follows from (8.47) that there exists a single v, denoted v^*, such that $\mathring{v} > 0$ (v increasing) if $\mathring{v} < v^*$ and $\mathring{v} < 0$ (v decreasing) if $v > v^*$. Hence $v(t) \rightarrow v^*(t)$, where from $\mathring{v} = 0$ we have that

$$\frac{s}{v^*} + \mathring{A} - n - \mathring{B} = \frac{\mathring{A}}{1 - \pi}$$

Now

$$\mathring{\pi} = (1 - \pi) \frac{\sigma - 1}{\sigma} \left(\frac{s}{v} + \mathring{A} - n - \mathring{B}\right)$$

Therefore in the neighbourhood of $v^*(t)$,

$$\mathring{\pi} = \frac{\sigma - 1}{\sigma} \mathring{A} \tag{8.48}$$

This relation supplements (8.46). The key role of the unitary value of the capital-to-labour substitution elasticity is evident. If capital can be substituted for labour relatively easily ($\sigma > 1$), then the share of capital costs would be rising and therefore the incentive to assimilate capital-saving inventions would be increasing. However, if $\sigma < 1$, the optimal bias moves from D towards C in Figure 8.3 until, in the limit, it becomes purely labour-saving. It is also only in this case that a growth equilibrium, one in which $v^* = $ constant, is globally stable (Drandakis and Phelps 1966).

Despite its strong assumptions, the theory is elegant and well rooted in the neoclassical tradition. It also answers our initial question. Unfortunately, it does so only conditionally. In order to serve as an explanatory theory of the labour-saving bias, it must assume that $\sigma < 1$ and regard the latter as a property of nature – an artefact. It must also assume that a stable Kennedy-type IPF, apparently unrelated to research inputs, exists and that firms select innovations in such a (myopic) manner as to maximize cost reduction only in the short term.

Localized search and a case for low capital–labour substitution elasticity

Innovation is a form of investment. As such, it involves costs and benefits that are usually spread over a number of years. Therefore inventive activity and the innovational decision must both be guided partly by the anticipation of future prices. If these are difficult to predict, as is often the case, then rather than the maximization principle various rules of an adaptive kind might be followed instead. A particularly appealing rule to adopt is to search for improvements in the neighbourhood of the techniques which are currently in operation. The production of existing goods and the operation of existing processes induce innovations in two ways. They generate new problems to the solution of which research effort would be attracted, and they stimulate the imagination of the researcher and instigate new ideas. In both cases, the inventions that may come about would very probably be close relatives, or improvements, of the currently employed innovations, rather than radically different solutions.

In terms of the aggregate production function, the inventive activity, according to this 'new view' of innovation, is concentrated in the neighbourhood of the point that corresponds to the set of production processes which the firms currently operate. Atkinson and Stiglitz (1969) call this form of inventive activity the localized search. The reader will note that this is in fact the same concept which is central in the Nelson–Winter evolutionary model, which has been discussed in Chapter 6. Our purpose now is to indicate how this alternative view, by virtue of its being capable of accounting for low capital–labour substitution elasticity, can be combined with an analysis of the Kennedy–Drandakis–Phelps type into a more plausible explanation of Harrod neutrality.

While innovation activity of the localized type may also result in a general shift upwards of the production function, such a shift would be expected to be particularly pronounced in the neighbourhood of the observed input-to-output ratios. This can be depicted graphically as in Figure 8.4. The relation between L/Y and K/Y is represented in Figure 8.4(b) by an isoquant. The curvature of the isoquant is indicated by the value of the capital–labour substitution elasticity; the higher the curvature at a given point of the isoquant the less easy it is to substitute capital for labour, and so the smaller

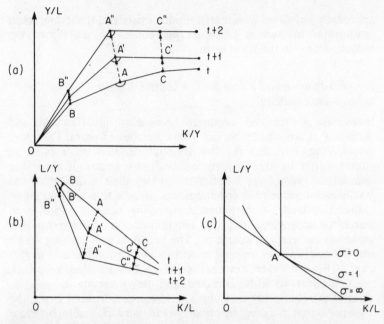

Figure 8.4 Isoquant curves (a) and (b) shift most in the neighbourhood of the techniques actually used, reducing the local elasticity of capital-labour substitution.

is the point substitution elasticity (see Figure 8.4(c), point A). From Figure 8.4 it is clear that localized technological change is likely to lead to reductions in the elasticity of substitution at the observed point. Atkinson and Stiglitz note that 'in some industries the effect of localization has been so strong that the advanced techniques (those adopted by the advanced countries) dominate the less-capital intensive ones, requiring both less labour and less capital' (Atkinson and Stiglitz 1969: 576). Even if examples of such dominance are rare, reducing the input-to-output ratios, above all in the neighbourhood of the techniques in use, is bound to result in a low elasticity of substitution between the inputs actually used.

We have noted in the previous section that a theory of the IPF type has to assume that $\sigma < 1$ if it is to account for the tendency of technological progress towards Harrod neutrality. The 'new view' theory of innovation would appear to provide a justification for this assumption. However, can the latter theory be regarded merely as improvement of the former theory in which the Drandakis–Phelps result also holds?

146

When technological progress is localized, the augmentation rates A and B can no longer be assumed to be functions of time alone. One way of formalizing this new approach is to assume, instead of (8.45), that

$$\mathring{A} = h(k/k^*, \mathring{B}) \tag{8.49}$$

where k is any capital-to-labour ratio, while k^* is the observed ratio. By (8.49), the economy is facing a different trade-off between $\mathring{A}$ and $\mathring{B}$ for each k/k^*, rather than a single frontier. It must further be assumed that for enterprise managers the only interesting trade-off is the one that corresponds to $k = k^*$, i.e. the localized innovation possibility frontier (LIPF). If this frontier is stable, the Drandakis–Phelps analysis may still hold. Technological progress need not be assumed to be purely factor augmenting for all k, but only at $k = k^*$ and near to it, where it matters. As the innovation effort is presumed to be concentrated in the neighbourhood of $k^*(t)$, we would expect the function h in (8.49) to have a form such that the local capital–labour elasticity is less than unity, with the consequence that the optimal bias is moving towards Harrod neutrality.

From $B^{HA} = (1 - \pi)B^{HI} - (1 - \sigma)\lambda$ it follows that, when $\sigma < 1$ and $B^{HA} = 0$, $B^{HI} > 0$, implying that, in the neighbourhood of the observed capital-to-labour ratio, technological progress would be capital-using in the Hicks sense.

We can interpret this result in yet another way. When a (more or less) constant fraction of output is earmarked for investment, capital stock increases faster than labour. The R&D staff is therefore permitted, indeed encouraged, to give attention and resources to developing processes that may involve an increase rather than a decline in k. Consequently, the rate λ is likely to be an increasing function of k in the neighbourhood of the observed capital-to-labour ratio k^*, implying the presence at that point of a capital-using bias in the Hicks sense ($B^{HI} > 0$). Since $B^{HI} = (\sigma - 1)(\mathring{A} - \mathring{B})$, $\sigma > 1$ would, in turn, imply $\mathring{B} > \mathring{A}$. Thus the Drandakis–Phelps analysis has the further implication that, with time, $\mathring{A}$ falls to zero while $\mathring{B}$ increases.

The discussion above is admittedly suggestive rather than definitive. In particular, the links between the specification of the IPF function, the value of the substitution elasticity on the actual growth path, and the dynamics of the whole system need fuller exploration.

The *n*-sector case

Concluding this chapter, let us return to the question of balanced growth in the context of Kennedy's *n*-sector model. Suppose that it is true that all technological change is localized and that all the substitution elasticities between capital inputs and labour are (significantly) less than unity. At the same time we continue to assume that in the neighbourhood of the actual growth path the technological progress observed is purely factor augmenting and the LIPF is of the Kennedy type:

$$\mathring{B}_j = L^j(\mathring{A}_{ij}, \mathring{A}_{ij}, \ldots, \mathring{A}_{n-1,j}) \tag{8.50}$$

Under these assumptions the Drandakis–Phelps result should apply to each of the *n* sectors, in which case technological change would tend to be Harrod neutral rather than Harrod–Kennedy neutral, implying that all $\mathring{A}_{ij} = 0$. Economic growth would continue to be balanced in the Kennedy sense only in the rather exceptional situation when $\mathring{B}_i = \mathring{B}_n$ for all *i*. In this case the composition of the economy would remain constant in both physical and value terms. Also constant would be all (physical and value) capital-to-output ratios, while labour productivity would be increasing at the same (constant) growth rate in all sectors.

However, even if the $\mathring{A}_{ij}$ are actually close to zero for all *i, j*, the observed labour productivity growth tends to vary considerably among sectors, indicating that, in fact, the rate $\mathring{B}_i$ varies among sectors. The conclusion of this analysis is that there is probably a tendency in each sector towards saving above all non-producible inputs, that the substitution elasticities may be rather low, and that fairly stable capital output ratios (but not sectoral composition) would in this case be consistent with significant variation in sectoral innovation rates.

Chapter nine

Variation of innovation rates among countries and over time: the First Hat-Shape Relationship

In this chapter the long-term rate of qualitative change continues to be the key economic variable to be explained. Although this rate is not observed directly, we have noted that under certain conditions the trend growth rate of output (or consumption) per man-hour is its approximate indirect measure, and that this trend growth rate can also be taken to be a proxy for the rate of technological innovation itself.

The historical data on output per man-hour are usually estimates of highly uncertain accuracy. Nevertheless it is fairly safe to say that in Western Europe and North America the rate of innovation, measured in terms of this proxy, had rarely been higher than 10 per cent per century until 1780, but since then it has rarely been less than 10 per cent per decade. In the course of the last two centuries, faster qualitative changes have begun to take place in other regions of the world as well. For this reason the two-century period is often said to be one of self-sustained and spreading 'technological revolution', a phenomenon that is unique in its intensity, brief by historical standards, and still continuing.

In the following chapter we shall attempt to develop a theory of innovation and growth in the course of that technological revolution. Before proceeding to do so, it may be instructive to look more closely at the process of economic growth in this period, particularly in more recent times, in terms of the variation over time and among countries of the major growth rates and ratios.

Stylized facts of macroeconomic history in the course of global industrialization

The economist who has done perhaps more than any other to document modern economic growth is Simon Kuznets. In support of the view that the period of technological revolution marks a *distinct economic epoch* Kuznets (1973) lists and discusses 'six

149

characteristics', of which two relate to aggregate growth rates, two others to structural transformation, and the remaining two to international spread. These are as follows.

(i) The growth rate of per capita product and of population are both 'large multiples' of the previously observable rates, especially in the developed countries.

(ii) The same is true for the rate of rise in productivity, whether measured in terms of output per unit of all inputs or of labour only.

(iii) A shift away from agriculture to non-agricultural pursuits is rapid. In the most advanced countries this has recently been followed by a shift away from industry to services.

(iv) Consequently, the process of modernization, in particular urbanization and secularization, is fast.

(v) The process towards a more integrated world economy has also quickened, with national exports now accounting for significant proportions of national outputs.

(vi) However, the spread of technology and economic growth to less developed countries is relatively slow. Consequently, the income gap, both absolute and (in many cases) relative, between the rich and poor nations is wider than ever before.

These characteristics relate mostly to things which are quantities. However, the important statements that they make about these quantities are qualitative in type and somewhat vague in content. Can we deduce something more definite and sharper about the nature of modern economic growth from the observed data?

The economist who suggested another list of six characteristics is Nicholas Kaldor (1961). His 'stylized facts' were assumed to be those the explanation of which, in his view, the modern theory of growth should regard as its primary task. They were intended to capture the essence of the growth experience of advanced capitalist countries only. However, factors related to the capitalist system and the high level of development actually play little role in Kaldor's characterization. His summary of history can therefore be thought to have wider applicability. His stylized facts relate primarily to aggregate (economy-wide) variables, but in part also to those concerning major sectors. They are as follows (Kaldor 1961: 177–9).

(i) Output and labour productivity increase at a steady trend rate.

(ii) Capital per worker also increases at a steady trend rate.

(iii) The rate of profit on capital is approximately constant.

(iv) The capital-to-output ratio over long periods is also approximately constant 'if differences in the degree of utilization of
capacity are allowed for'.
(v) There is a long-term steadiness in the share of profits and in
the investment-to-output ratio, and a high positive correlation
between the two.
(vi) While these constancies relate to particular economies in their
growth over time, there is an 'appreciable variation in the rate
of growth of output and labour productivity among different
economies at any one time and a corresponding variation in
the investment ratio and the profit share'.

Except for (vi), the characteristics above are seen to make much
more definitive assertions about the magnitude of key growth rates
and parameters than those of Kuznets. However, are these
assertions acceptable as approximations? If not, what amendments
need to be made? We shall first attempt to answer these two
questions, and then add a new characteristic, intended to give
more substance to Kaldor's fact (vi).

Thanks to a series of studies of historical growth statistics,
summarized and extended by, among others, Angus Maddison
(1979, 1980) and Walt Rostow (1978), we are now in a better
position than Kaldor was in the late 1950s to judge the stability of
aggregate growth rates and major ratios. The proportional change
in the GNP per man-hour in sixteen advanced capitalist countries
in the course of the last century has been (remarkably) stable only
in the United States (Maddison 1979: 43, Table 1 and Annex). In
the other fifteen countries there has been a sharp difference
between the period 1870–1950 and the years 1951–90. Clearly
there must have been something important which produced this
difference and which Kaldor, in 1961, might have ignored as
merely an expression of post-war reconstruction.

Note that in (vi) Kaldor acknowledges the presence of considerable variation in the productivity growth rate between countries.
Yet this variation is apparently assumed to have no effect on the
presumed 'steadiness' of the productivity and output growth rates
in individual countries. Thus the Kaldorian world behaves as if it
were a collection of mutually independent economies, each
moving along its own balanced growth path. In such a world,
technological and (therefore) productivity gaps would have to
arise, since the (constant) productivity growth rates differ. In many
countries these gaps have grown to become eventually very large.
However, they are not supposed to correct or in any other way
influence the divergent balanced-growth paths of the countries, as

151

if the gaps were assumed by Kaldor to have no feedback effect on the innovation rates.

Since the time Kaldor first formulated his stylized facts, it has become apparent that the assumption is false. The productivity growth data of the post-1950 years clearly indicate that the effect has been strong enough to produce a causal relation, which is our hat-shape relationship, between the productivity growth rates of the countries and the intercountry productivity gaps (Gomulka 1971). However, has the relationship been in evidence at earlier times as well?

In 1870 the United Kingdom was the 'technologically leading' country with a substantial productivity lead over most European countries and Japan, as well as a 20 per cent lead over the United States (Maddison 1979: Table 2). Among the sixteen countries the UK economy was also the largest national economy. Although soon to be overtaken by the US economy, in terms of both productivity and size, the UK economy was by far the largest exporter of manufactured goods throughout the period 1870–1913 and hence the prime source of technological influence for the smaller European economies. This role has been played by the US economy only since the Second World War, especially in the 1950s and 1960s.

Maddison's data can be used to estimate the relation between the rate of labour productivity growth and the productivity gap. The results are as follows (Gomulka and Schaffer 1987):

$$\mathring{y}_{i,1870-1913} = 1.47 + 0.41 \log\left(\frac{y_{UK}}{y_i}\right)_{1890} \qquad R^2 = 0.156 \qquad (9.1)$$
$$\phantom{\mathring{y}_{i,1870-1913} = } (0.14) \ (0.21)$$

$$\mathring{y}_{i,1950-70} = 2.16 + 3.48 \log\left(\frac{y_{USA}}{y_i}\right)_{1960} \qquad R^2 = 0.731 \qquad (9.2)$$
$$\phantom{\mathring{y}_{i,1950-70} = } (0.40) \ (0.54)$$

where $\mathring{y}$ stands for the growth rate of y, which is the GDP per man-hour, and the standard errors of the estimated coefficients are in parentheses beneath the respective values. Note that, since productivities tend to rise exponentially, $\log(y_{UK}/y_i)$ in (9.1) equals the trend growth rate of y_{UK} multiplied by the number of years needed for country i to reach the UK level. The same interpretation applies for the $\log(y_{USA}/y_i)$ in (9.2).

In the two subperiods, the gap coefficient b is seen to be positive and statistically significant at the 95 per cent probability level. It is interesting that the coefficient is much higher for recent years than for the period 1870–1913. This result can be interpreted to reflect a number of developments relating to human, institutional, and

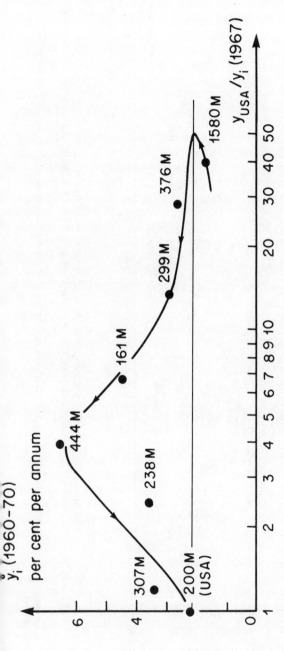

Figure 9.1 The Hat-Shaped Relationship when countries are grouped according to their gross national product per capita. The black spots represent the appropriate groups of countries, the numbers giving their combined population. Note that y_i is the per capita gross domestic product in country i, $\dot{y}_i$ is its growth rate, and y_{USA} is the per capita gross domestic product in the United States.

Source: The data used are those reported in Rostow, W.W. (1978) *The World Economy: History and Prospects*, London: Macmillan, pp. 564–5, Table V-84

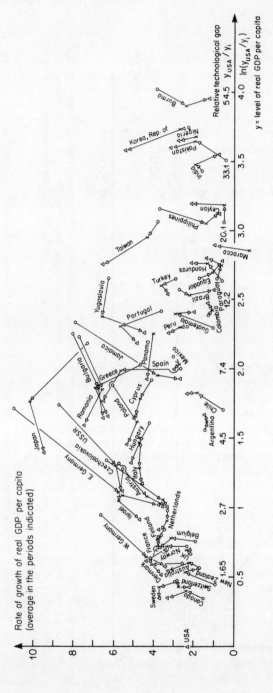

Figure 9.2   The Hat-Shaped Relationship when countries and their growth paths are shown individually: ○, (1953, 1950–60); ×, (1960, 1955–65); △, (1965, 1960–8). The same variables as in Figure 9.1 but in relation to specific countries.

Source: Gomulka, S. (1971) *Inventive Activity Diffusion and the Stages of Economic Growth*, Aarhus: Aarhus University Press, p. 66, Figure 4.2

technological capabilities which favoured more rapid transfer of foreign technology and its faster diffusion within Europe and Japan after the Second World War.

The evidence above, strictly speaking, is only for the presence of a positive association between the productivity gap and the rate of productivity change; no causality is implied. However, the evidence can be interpreted as consistent with the notion that the gap is an aggregate indicator of some of the factors that influence the rate of international technological transfer and hence the growth rates of innovation and labour productivity in the technology-importing countries.

This interpretation, it should be noticed, need not be put in doubt on the grounds that the cross-country variation in $\hat{y}$ can in part be explained by the corresponding variation in the growth rate of the ratio $k = K/L$ and that, in relations (9.1) and (9.2), the gap-related term may merely reflect, partly or wholly, the productivity effect of the substitution of capital for labour. This would have been a valid point if the growth rates refer to short or medium periods of time, in which the causality may justifiably be thought to run from $\hat{k}$ to $\hat{y}$. However, the growth rates in the relations above are 'long term'. In such relations the capital stock must be treated as a reproducible input, the growth of which is dependent, in the first instance, on the growth of output and ultimately on the rate of innovation (as well as the growth rate of employment). When each country's capital-to-output ratio is the same at the beginning and the end of any such (long) period, so that $\hat{k} = \hat{y}$, then the cross-country variation in both $\hat{k}$ and $\hat{y}$ would have to be explained by the variation among the countries in the rate of innovation. The gap-related term in (9.1) and (9.2) simply captures both the direct effect and all the indirect effects of the international transfer of technology. However, the critical assumption needed for this interpretation is that the productivity effects of any actual variation in the capital-to-output ratio, not to mention the aggregation-type and other problems, can be ignored (see Chapters 1 and 8).

It is interesting that the US labour productivity growth rate continued to exceed the UK growth rate for some 60 years *after* the US productivity level surpassed the UK level at about 1890. This is a separate issue, but it may be useful to comment on it briefly, as the phenomenon can be taken as evidence against the feedback effect hypothesis. According to Maddison (1979: 17):

The USA developed its productivity lead initially in the period from the 1890s to 1913 at a time when its prospects were particularly bright because of its great natural resource advantages, huge internal markets, and rapid population growth. This fostered higher rates of investment than in Europe and a faster growth of capital per employee.

Indeed, the ratio of gross investment to GDP was at that time about 20–25 per cent in the United States and only 10–15 per cent in the United Kingdom. However, the difficulty with Maddison's argument is that a higher US investment ratio may have been caused by, rather than itself have caused, a higher US innovation rate and a higher rate of population growth. The reason is that a high innovation rate is needed to generate profitable investment opportunities and thus sustain a faster growth of capital per employee over a long period of time. The underlying reason for the US productivity overtaking the UK level is therefore likely to lie elsewhere, namely in the fact that the centre of gravity of the world inventive and innovative activity was then shifting from the United Kingdom to the United States. The size of inventive activity in the two countries at that time is difficult to judge, but it can be gauged by the size of their industrial sectors. As it happens, it was also in the 1890s that the volume of the US industrial output surpassed that of the United Kingdom. Twenty years later the United States accounted for a third of world industrial output, more than twice the United Kingdom's share (Rostow 1978: 52–3). It was this faster US quantitative expansion, helped immensely no doubt by a large immigration and brain drain from Europe as well as by locally abundant natural resources, that probably enabled a similarly faster growth of US inventive activity and hence also enabled a faster qualitative change. The larger size of the US economy, at a time when Europe had not yet made much progress in integrating its national economies, would also have offered more market opportunities to make immediate use of the inventions designed to exploit large-scale economies, especially in the mass-production consumer goods industries.

What has happened in the period since 1913 is described by Maddison in the following way:

> By 1913 the US productivity advantage over the UK – the old leader – was about a quarter. One cannot tell how wide this productivity gap would have become in 'normal' circumstances. Eventually the forces making for US ascendancy would have faded, as indeed they now have. In the meantime, however, the productivity gap became very much bigger mainly because of the two world wars – both of which stimulated the US economy

and retarded the advantage of the other countries.

In 1950, there was an unnatural degree of dispersion between the USA and most of the other countries. This was why the latter did so well in growth terms simply by implementing sensible policies for full employment and freer trade. High demand propelled them into a situation of unprecedently high investment and eliminated a good deal of their technical backlog. (Maddison 1979: 17)

A similar interpretation of the strong post-war upswing in European and Japanese productivity growth has been given by Abramovitz (1979), as well as by some other writers, including Gomulka (1971, 1979), all of whom emphasize a number of developments relating to human, institutional, and technological capabilities which favoured more rapid transfer of foreign technology and its faster diffusion within Europe and Japan after the Second World War.

Let us now consider Kaldor's other stylized facts. First of all we note that (ii) and (iii) are not independent characteristics, but implications of (i), (iv), and (v). If capital-to-output ratio is constant, then, by characteristic (iv), capital and output grow at a common rate. In view of (i), the capital per worker must then grow at the same constant rate as does the output per worker (characteristic (ii)). The constancy of the profit-to-capital ratio (fact (iii)) follows, in turn, from the presumed constancy of the profit-to-output ratio and the capital-to-output ratio. (Profit can be defined in a number of ways, and the historical behaviour of the various profit-to-output ratios in the United Kingdom is discussed well by Graham Haache (1979: Chapter 15).) On the whole income shares do tend to be relatively stable, although there seems to be a long-term tendency for the profit share to decline and the income share of the employees to increase. These tendencies possibly mirror partly the diffusion of capital ownership within modern societies and partly the increasing use by governments of tax revenues for investment purposes. In any case we shall make no particular use of the alleged constancy of the profit share in the following pages.

The steadiness of the aggregate (or sectoral) capital-to-output ratio, both in current and constant prices, is the characteristic that we have already invoked for two quite different reasons. First, when the capital-to-output ratio (aggregate or sectoral) is constant, the contribution of all qualitative changes to economic growth, over a long period of time and under constant returns to scale, is measured by the rate of growth of labour productivity. Since the latter is observed, this first reason relates to measurement convenience. Second, if the capital-to-labour substitution elasticity is less than unity, as the 'new view' of technological change's being largely localized indicates to be the case in practice, then innovations are likely to be predominantly Hicks labour-saving (or Harrod neutral). This type of technological change happens to produce a growth path along which the capital-to-output ratio does tend to be stable. However, if technological progress is strongly Harrod non-neutral, the capital-to-output ratio would be unlikely to be stable. Thus the observed stability of the ratio can be taken as evidence in support of the view that the localized possibility innovation frontier, the innovation-inducing role of prices, and local search are not merely interesting theoretical concepts but also capture some essential features of economic reality.

The data on the stock of capital are generally of poorer quality and more difficult to come by than those on output. However, the data, such as they are, do suggest that, *while certainly not a constant, the aggregate capital-to-output ratio has been substantially more stable than the aggregate labour-to-output ratio.*

Thus the US aggregate K/Y was rising from the year 1880 until 1920, but has since been declining to reach approximately the 1880 level a century later. According to one source, 'only some 17% of the rise in 1880–1919 is attributable to changing industrial composition' (Domar 1961b: 106). The sectoral capital-to-output ratios have tended to follow a similar pattern. For instance, in US manufacturing the ratio increased from 1.5 in 1880 to 2.6 in 1919, but then declined to 1.6 in 1948.

In Germany the aggregate K/Y stayed at about 4 throughout the period 1850–1950 (Hoffman 1961: 106). In the United Kingdom, however, the aggregate K/Y declined from 5.1 to 3.7 in the years 1855–1913, and then increased to 4.6 in 1976 (Haache 1979: 274). A similarly stable pattern of capital-to-output ratios, for both total economies and major sectors, was also present in the USSR and East European countries in the period 1950–1980, although the ratios tended to decline in the 1950s and to rise from the second half of the 1970s, with the pattern reflecting changes in the utilization of capital (the utilization rose in the years 1945–60 but fell in the 1970s and the 1980s), an upward trend in the share of capital-intensive investments in agriculture and infrastructure, and, in the case of the USSR, a shift to regions such as Soviet Siberia.

To conclude this venture into the territory of historical statistics, it is probably fair to say that Kaldor's steady state characterization of economic growth is good only for studies of individual countries both in isolation from other countries and within periods of time that can be viewed as distinct phases of each country's development. Between such phases the trend growth rates of output and labour productivity may undergo substantial changes. In contrast with growth rates, the trend capital-to-output ratios for major sectors and whole economies tend to be (remarkably) stable, but not necessarily so within particular phases. It follows that, for example, a rise in the trend growth of output from one phase to another would tend to be associated with a rise in the investment-to-output ratio rather than a fall in the capital-to-output ratio.

The capital-to-output ratios are also found to be very diverse among individual branches. The US studies by Leontief and his group, based on a 68-industry classification for 1939 and a 192-industry classification for 1947, reveal a range of capital-to-output coefficients from about 0.1 to 10 (after Domar 1961b: 103). These less aggregated coefficients are also more variable over time, though again the changes – unlike those of the corresponding labour-to-output ratios – are not all in one direction.

Catching-up and the First Hat-Shape Relationship: a preliminary interpretation

The search for a pattern in the observed wide variation in the cross-country growth rate of output per man-hour has, as we have already noted, led to the observation that the latecomers in industrialization should, and in fact do, tend to innovate faster than the world's 'technology frontier area' (TFA), which is defined as the regions in which the world's best technology is employed. The reason behind this observation is the common experience that in technology or organization, as well as in science, learning and imitating is typically cheaper and faster than is the original discovery and testing. In the preceding section we have taken the distance between the level of development of the TFA and that of a less developed country (LDC) as a measure of the backlog of technological opportunities to exploit. The larger is the backlog, the greater can we expect to be the economic incentive to take advantage of some of these opportunities and, other things being equal, the greater the rate of international technology transfer.

The idea that there might be 'advantages of backwardness' in this sense is usually associated with the names of Thorstein Veblen and Alexander Gerschenkron. Veblen (1915) applied it to Germany *vis-à-vis* England, and Gerschenkron (1962) updated and extended the work to include Russia, France, and Italy. In a

159

formalization of this idea Nelson and Phelps (1966) assumed that an increase in the level of technology of an LDC, to be denoted by T, is proportional to the technology $T^* - T$ between the LDC and the TFA, i.e.

$$\dot{T} = \beta(T^* - T) \tag{9.3}$$

On dividing through by T, it follows from (9.3) that

$$\frac{\dot{T}}{T} \equiv \alpha = \beta\left(\frac{T^*}{T} - 1\right) \tag{9.4}$$

Thus the Nelson–Phelps assumption implies that the relationship between the rate of innovation and the relative technology gap is, for any LDC and in the course of time, positive and linear. Moreover, the LDC's innovation rate α would always exceed the innovation rate α^* of the TFA but fall toward it asymptotically, with the relative gap falling as a result towards a country-specific positive constant called the 'equilibrium (relative) technology gap'. This reduction in the relative technology gap between an LDC and the TFA is what is meant by (international and/or technological) *catching-up*.

The equilibrium gap can be obtained from (9.4) by substituting α^* for α. Thus T/T^* increases at most to a level $1/(1+\alpha^*/\beta)$. It follows that the catching-up is more effective the lower is α^*, or the speed of the moving target, and the greater is β, a parameter reflecting the absorptive capacity. It may be noted that β is assumed to be independent of the level of development. It is this evidently unrealistic assumption which implies that the least developed countries should be also the most innovative. The world pattern of productivity growth rates observed in the period 1950–90 suggests the need for a modification of the original Veblen–Gerschenkron hypothesis. Namely, for the group of highly backward LDCs backwardness is a disadvantage, with the rate of innovation tending to be lower the greater the relative technology gap. The relationship across all countries is thus of the 'hat-shaped type' (Gomulka 1971; Horvat 1974).

The usual interpretation of the negative part of the hat-shape relationship rests on the concept that the severely limiting factor in the initial phase of the catching-up is 'absorptive capacity'. As educational standards and physical infrastructure are improved, a larger amount of foreign technology becomes profitable. Technology imports themselves also help upgrade skills and increase exports, attracting still larger technology imports, and so forth. It

160

is, essentially, this causality sequence which gives rise to the relationship's negative part. An implication of this modified Veblen–Gerschenkron hypothesis is that, before absorptive capacity is developed to reach a level at which an LDC's rate of innovation is the same as that of the TFA, an LDC's relative backwardness would be increasing.

Formal analysis of the hat-shape relationship is given in Chapters 11 and 12, where we shall interpret it as an international macroeconomic equivalent of logistic or S-shaped diffusion curves usually observed for individual inventions. Theoretical studies model the dynamics of catching-up under different channels of technology transfer, such as direct foreign investment (Findlay 1978), cost-free diffusion (Gomulka 1971), or trading conventional goods for disembodied technology (Gomulka 1970). In our analysis, however, we shall also take into account, and indeed emphasize, economic dualism and technology transfer costs.

Empirical studies appear to indicate that embodied technology transfer is an important, perhaps the main, channel for most LDCs. However, in the post-war catching-up of the United States by countries with large R&D sectors, such as Japan, the FRG, and the USSR, the import of capital goods from the United States has apparently played a small role, suggesting that disembodied diffusion, both (virtually) cost free and commercial, has probably played the main role. The post-1975 labour productivity slowdown in these countries can be interpreted as evidence of the countries approaching their specific equilibrium technology gaps. These equilibrium gaps, as well as innovation rates in the course of the catching-up itself, are apt to be influenced by cultural and systemic factors. The process of catching-up is therefore bringing about a state of international growth equilibrium in which the innovation rate would be approximately common to all countries but in which productivity and technology levels would continue to vary among the world's countries, although to a much lesser extent than at present. How these equilibrium gaps can be estimated is the subject of the next section.

Catching-up by the OECD and CMEA countries

We begin by estimating the relationship between what we shall call the trend rate of innovation and the relative technological gap. While in general of the hat-shape type, this relationship for the fairly advanced countries which are of concern here is positive, or of the standard Veblen–Gerschenkron type. One such relationship was estimated for the industrial sector of the seven Council of

Macroeconomics of innovation

Mutual Economic Assistance (CMEA) countries: Bulgaria, Czechoslovakia, GDR, Hungary, Poland, Romania, and the USSR, to be referred to below as the Seven. Another relationship was estimated for the manufacturing sector of the twelve Organization of Economic Co-operation and Development (OECD) countries: Canada, Finland, France, the FRG, Greece, Ireland, Italy, Japan, Norway, Sweden, the United Kingdom and the United States, to be referred to below as the Twelve. Both relationships were estimated on the basis of data for the period 1955–83 (Gomulka and Schaffer 1987; Gomulka 1988). The systemic (as well as cultural and other) factors can be thought to underlie the difference between these relationships. This is thus one way of identifying the impact of such factors.

For the purpose of identifying the trend rate of innovation it is useful to recall Kalecki's equation for the growth rate of output in an economy in which fixed capital is of the putty-clay (vintage) type:

$$\overset{\circ}{Y} = \frac{1}{m} \frac{I}{Y} - a + u_1 \tag{9.5}$$

where a circle above a symbol denotes growth rate, m is the average capital-to-output ratio for new investment projects at their full utilization, I is the volume of all these projects, a is the fraction of total output lost due to wear and tear, and u_1 is the fraction of total output which is either gained through organizational improvements or lost if 'barriers' to growth prevent full capital utilization. Since equipment needs to be manned, we also have a second growth equation which links output and employment. Assuming constant returns to scale, this equation has the form

$$\overset{\circ}{Y} = \overset{\circ}{L} + \alpha + \text{substitution term} + u_2 \tag{9.6}$$

where $\overset{\circ}{L}$ is the growth rate of employment, α is the Harrod–Kalecki rate of innovation, u_2 is the term reflecting changes in the rate of labour utilization, and the substitution term is zero if the capital-to-output ratio is constant. If labour rather than capital is the given growth-constraining factor, then (9.6) determines $\overset{\circ}{Y}$ and (9.5) is an equation for I/Y, given m, a, and u_1. However, if a third factor, such as foreign exchange or energy, forces a slowdown in $\overset{\circ}{Y}$, then, given I/Y and $\overset{\circ}{L}$, the term u_1 falls and u_2 becomes negative. In this case we would expect a downward adjustment in the investment ratio I/Y and in the growth rate of employment $\overset{\circ}{L}$, so that the utilization rates for capital and labour can return to their 'normal' levels. By the trend growth path we

162

mean a path along which these two rates are maintained at such normal levels, i.e. when u_1 is any constant and $u_2 = 0$. On such a path, the key parameter determining the growth rate of output is a, while u_1, given m and a, influences only the investment ratio.

Equation (9.6) also holds in the standard approach in which all capital is technologically homogeneous. In this case $\mathring{Y} = \eta_K \mathring{K} + \eta_L \mathring{L} + \lambda + u$, where η_K and η_L are the capital and labour elasticities, λ is the Hicks rate of innovation, and u is a term related to changes in the utilization rates of capital and labour. Subtracting $\eta_K Y$ on both sides and dividing through by $1 - \eta_K$ gives our standard equation (Chapter 1)

$$\mathring{Y} = \frac{\eta_K}{1 - \eta_K} (\mathring{K} - \mathring{Y}) + \frac{\eta_L}{1 - \eta_K} \mathring{L} + \frac{\lambda + u}{1 - \eta_K} \tag{9.7}$$

The first term on the right-hand side is the Harrod–Kalecki substitution term, which is zero when the K/Y ratio is constant. When $1 - \eta_K = \eta_L$ (returns to scale constant), then (9.7) must coincide with (9.6) so that $\lambda/(1 - \eta_K) = a$. We regard a as a measure of the trend rate of innovation, as it is the contribution to growth of all the qualitative changes (innovation) when $\mathring{K} = \mathring{Y}$, so that K/Y is constant. The unexplained residual includes $u/(1 - \eta_K)$ as well as any error term.

The trend innovation rate, catching-up and equilibrium gap: CMEA and OECD, 1955–85

An estimate of the trend innovation rate depends in part on the purchasing power parities (PPPs) used. Gomulka and Schaffer (1987) use two sets of parities for the CMEA countries, one due to Alton and the other due to Heston and Summers, and one set for the OECD countries, derived by the authors from the PPPs due to Krevis, Summers and Heston (1984). When official data on growth rates are used, the industrial production function estimated for the two groups of countries, separately as well as jointly, is of the following type:

$$\mathring{y}_t = \eta_K \mathring{k}_t + s\mathring{L}_t + \gamma \ln x_t + \delta\lambda_1 + \text{unexplained residual} \tag{9.8}$$

where y is the value added per man (or per man-hour where data are available), k is the capital-to-labour ratio, x is the ratio of y in the United States to that in any specific country, taken in the

163

middle of period *t*, and δ is a dummy variable equal to unity for the Seven and zero for the Twelve. The production function is thus of the Cobb–Douglas type, in which $s = \eta_K + \eta_L - 1$, and $\lambda_t = \gamma \ln x_t + \lambda_0 + \delta\lambda_t$. A positive *s* implies that there are productivity gains to be made when employment increases. Gomulka and Schaffer (1987) interpreted these gains as being due to a greater scope for the growth of the technologically more advanced sector. Another possible interpretation of a positive *s* is that limited labour mobility is less (more) of a constraint to the growth of output when employment increases (declines). Both are cases of dynamic economies of scale. A variable component of technological change $\gamma \ln x_t$ is intended to capture the effect of international technology transfer from the TFA to a particular country, with the US manufacturing sector serving as a proxy for the TFA. As was noted in the previous section, ln *x* equals the trend growth rate of labour productivity in the US manufacturing sector, the rate to be denoted by α^*, multiplied by the number of years needed for any specific country to reach the US level of labour productivity. Thus *x* is a measure of the technology gap in terms of the number of years of lagging behind.

From (9.7) it follows that the trend rate of innovation α is the growth rate $\mathring{y}$ which obtains when $\mathring{k} = \mathring{y}$ and $\mathring{L} = 0$, which is therefore a rate that, as far as capital accumulation and employment are concerned, can be sustained over a long period of time. Given (9.8), an estimated trend rate of innovation has the following form:

$$\alpha_t \equiv \mathring{y}_t^{\text{trend}} = \frac{\gamma}{1 - \eta_K} \ln x_t + \frac{\lambda_0 + \delta\lambda_1}{1 - \eta_K} \tag{9.9}$$

This relationship between $\mathring{y}_t^{\text{trend}}$ and x_t can be estimated using pooled data or, alternatively, using the data for each of the two groups of countries separately. In the latter case the result is as shown in Figure 9.3. The systemic (and other 'non-economic') factors are seen to influence both the trend rate of innovation at any given level of x_t and the end result of the catching-up, which is the *equilibrium* productivity gap x^*. The latter can be calculated from (9.9) on substituting α^* for $\mathring{y}_t^{\text{trend}}$.

If we take Alton's PPPs and official growth rates, the CMEA's trend innovation rate is lower than that of the OECD almost uniformly by about 1.5 per cent. This estimate of the 'dynamic efficiency gap' would have to be increased, per cent for per cent, by any under-reported inflation. However, for Summers–Heston PPPs and official growth rates there is virtually no dynamic efficiency gap, but even in this case the equilibrium productivity

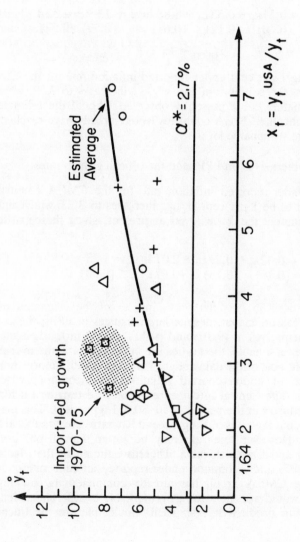

Figure 9.3 The 'comparable' relationships between labour productivity growth $\mathring{y}_t$ and relative technological gap x_t implied when taking, for all countries and all time, $\mathring{y}_t = \mathring{k}_t$ and $\mathring{L}_t = 0$: ○, (1955–60, 1957); +, (1960–5, 1962); △, (1965–70, 1967); □, (1970–5, 1972); ▽, (1975–83, 1980).

gap for the CMEA, at 64 per cent of the achievable productivity level, is fairly substantial.

If equation (9.8) is estimated on the basis of pooled data (twenty-four observations for the OECD countries and thirty-two observations for the CMEA countries), it is as follows (standard errors in parentheses).

(i) Alton's PPPs and the official growth rates:

$$\hat{y}_t = 0.31\mathring{k}_t + 0.33\mathring{L}_t + 1.88 \ln x_t + 1.89 - 1.52\delta \qquad (9.10)$$
$$\phantom{\hat{y}_t = }(0.10) \quad (0.12) \quad (0.61) \qquad (0.59) \ (0.57)$$

$$R^2 = 0.57 \qquad x^*_{\text{OECD}} = 1.02 \qquad x^*_{\text{CMEA}} = 1.67$$

Assuming 1 per cent under-reported inflation rate for the CMEA countries, the equilibrium gap x^*_{CMEA} increases to 2.32, which is approximately the x presently observed. Should the estimate be about right, the CMEA countries would already have reached the maximum sustainable level.

(ii) Summers–Heston PPPs and the official growth rates:

If the under-reported inflation rate for the CMEA countries is assumed to be 1 per cent, x^*_{CMEA} increases to 3.23, which again is approximately the x observed at present, given these particular PPPs.

$$\hat{y}_t = 0.32\mathring{k}_t + 0.29\mathring{L}_t + 2.07 \ln x_t + 1.73 - 0.95\delta \qquad (9.11)$$
$$\phantom{\hat{y}_t = }(0.10) \quad (0.13) \quad (0.67) \qquad (0.60) \ (0.51)$$

$$R^2 = 0.55 \qquad x^*_{\text{OECD}} = 1.05 \qquad x^*_{\text{CMEA}} = 1.67$$

It is reassuring that, despite large changes in PPPs, the coefficients themselves in (9.10) and (9.11) change little, their magnitudes are reasonable, the t ratios are fairly high, and the overall fit, given that most of the data are growth rates, is satisfactory. A 1 per cent rate of under-reported inflation does not appear to be excessive. The Central Intelligence Agency's estimate of it for the Soviet industry in the period 1950–80 is 1.6 per cent. This may be too high, but the rates of Bulgaria and Romania are possibly at least as high. However, the rate may be lower than 1 per cent in Hungary and Czechoslovakia. Therefore the result that, for both sets of PPPs and a common under-reported inflation rate of 1 per cent, the CMEA group has already completed its catching-up process would appear to be as remarkable as it is plausible. The equilibrium productivity gap itself should, in turn, be traced to

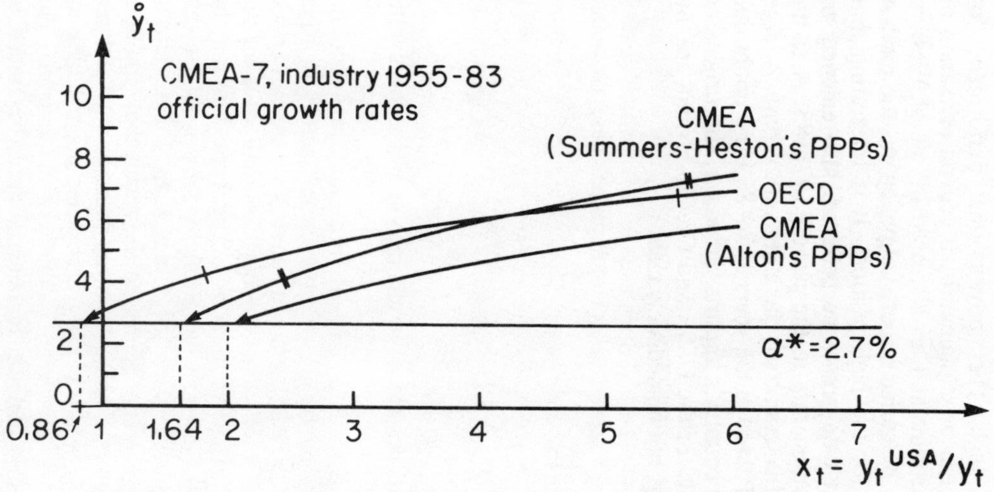
CMEA-7, industry 1955-83
official growth rates
CMEA
(Summers-Heston's PPPs)
OECD
CMEA
(Alton's PPPs)
α*=2.7%
10
8
6
4
2
0
0.86
1
1.64
2
3
4
5
6
7
$\mathring{y}_t$
$x_t = y_t^{USA}/y_t$

Alton's PPPs	$\mathring{y}_t = 3.0 \ln x_t + 0.6$
Summer-Heston's PPPs	$\mathring{y}_t = 3.8 \ln x_t + 0.8$
OECD-12, manufacturing 1955-83	$\mathring{y}_t = 2.3 \ln x_t + 3.0$

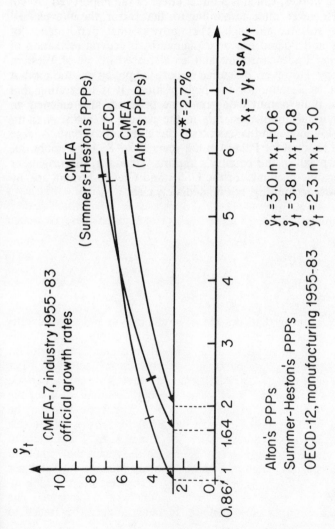

Figure 9.4 Actual labour productivity growth and the underlying relationship between $\mathring{y}_t$ and x_t, when $\mathring{y}_t = \mathring{k}_t$ and $g_L = 0$. For the industrial sectors of the Seven the observations are as follows (first x_t, second $\mathring{y}_t$).

various time-lags, which give rise to an equilibrium technological gap, and to 'normal' static inefficiency (overmanning and under-production). Gomulka (1986, 1988) has attempted to 'explain' the productivity gap for the USSR in this way.

Deviations from the trend innovation rate for the CMEA countries can be seen in Figure 9.4. They are particularly large for Bulgaria, Poland, and Hungary in the period 1971–5. The devi-ations are in part due to the unusually high growth rate of capital stock per worker, which is a direct effect of the import-led growth policy. However, after accounting for that factor, the unexplained residuals still remain high. The above-normal performance of Hungary and Poland can be explained by a general relaxation of the foreign trade constraint and an increased inflow of Western technology. However, Bulgarian import-led policy was too modest an affair to account for the high residual. It is interesting that Bulgaria, if its output data are to be believed, also enjoyed an 'above-normal' innovation rate in the period 1976–85. A sharp growth slow-down in that country is therefore yet to come. A large negative residual for Poland in the years 1976–83 is no doubt due to a sharp fall in the country's imports. However, the origins of large negative residuals for the USSR and Czechoslovakia are, by this growth accounting, not immediately clear.

Chapter ten

'Technological revolution' as an innovation superwave in the world technological frontier area

The 'world technological frontier' has been defined as comprising all production methods which at any given time are either the most economical or the most productive in the world. If they are known to firms, these world-best methods would be the only candidates for selection, the actual choice depending as usual on local prices, output demands, resource endowments, and so forth. There may also be firms which know only some of the methods in question but whose choices under this limited knowledge, would remain the same as when their knowledge was complete. The collection of firms of both categories is what we call the 'technological frontier area' (TFA).

It may be expected that at any one time only a small proportion of all the world's firms belong to the TFA and that the firms which do belong are distributed among national economies in a highly uneven manner, concentrated largely in countries where the income per head is at or near the world's highest. For the purposes in hand it will be useful to regard the TFA as a single economy. We shall also assume that the inventive activity of that imagined economy is the only factor capable of moving the frontier outwards. The inventive activity of the firms operating behind the frontier will thus be ignored. As an approximation we can think of the United Kingdom and parts of continental Europe as the TFA for most of the nineteenth century, and of the United States and parts of Western Europe as the TFA for most of the twentieth century.

An important model of an economy such as our TFA was developed by Edmund Phelps (1966). His two-sector model is representative of a family of simple macroeconomic models of innovation and growth, most of them created in the 1960s and 1970s. Their primary purpose was to find the rates of balanced growth of technology and output which would obtain when resources of labour and capital are distributed in an 'optimal' manner between the two sectors: conventional production and inventive activity.

However, the central feature of innovation and growth of the world economy in general and of its TFA in particular, during the last two centuries, has been an unbalanced growth: an expansion of the inventive activity much faster than that of conventional production. Two centuries is a relatively short period of time by historical standards. The unusually rapid innovation during that period is therefore still an exception in the history of humanity. Since the faster growth of the inventive activity must come to an end, probably in the course of the next century, there is a possibility that the rapid innovation could also come to an end. This innovation slow-down would have rendered the technological revolution to be at once a transitory phenomenon and one which would for ever be seen as a huge innovation outburst or an innovation superwave.

The primary purpose of this chapter is to investigate this prospect in terms, above all, of its economic theory in order to identify the key assumptions and parameters which are to affect its likelihood and the time-scale. The analysis represents a development of the ideas presented by Gomulka (1971). Since Phelps's model is a point of departure for this analysis, we begin by presenting it in some detail.

The Phelps model of innovation and balanced growth

The key equations of the model are as follows:

$$Y = F(K, TN) \tag{10.1}$$

$$\mathring{T} = H(E, T) \tag{10.2}$$

$$E = M^{\beta} R^{\mu} L^{\gamma} \tag{10.3}$$

$$L = N + R = L_0 \exp(nt) \tag{10.4}$$

According to (10.1), the net output Y of the conventional sector is dependent on the capital stock K and the 'effective' labour TN, where T represents an index of technology and N represents labour input in terms of man-hours. Technological change is thus assumed to be purely labour-saving. Equations (10.2) and (10.3) represent an embedded two-level production function for the technology sector. At the more basic level, E is the amount of research produced when researchers R are equipped with capital M and selected from a total labour force L. This research, in turn, brings new technology ΔT, and in (10.2) this addition is assumed to be influenced positively by T itself. The reason is that innovation also builds on accumulated past research. In (10.1) and (10.2) constant returns to scale are assumed, but the elasticities of substitution between K and N and between E and T need be neither unitary nor constant. In (10.3) the returns to scale need not be constant, but the elasticity of substitution between any two of the three 'factors' is unitary. The latter assumption is highly restrictive and we shall dispense with it later in the paper.

Let us denote the partial elasticity of the function F with respect to K by a, and the partial elasticity of the function H with respect to E by b. The assumption of constant returns to scale implies that $a = a(K/TN)$, $b = b(E/T)$, and $0 < a, b < 1$.

The Phelps technology production function has two intuitively appealing properties. One is that the same research effort will be more productive if it is spread evenly over a longer period of time rather than being concentrated in a short period. To see this, consider a variable period Δt, a steady research flow E, and a constant total research effort $E\Delta t$. Denoting the latter by c, we have that $E = c/\Delta t$ and $\Delta T = H(c/\Delta t, T) \Delta t$. Hence $\partial \Delta T/\partial \Delta t = (1 - b)\Delta T > 0$, which confirms that ΔT increases with Δt, given c. An implication of this property is that research effort allocated evenly over a period of time is assumed to be more productive than an equivalent total research effort which proceeds in fits and starts.

Another property of the model is that it attempts to capture, admittedly in a very crude manner, the inherent heterogeneity of people with respect to their inventive ability. If we assume that in the inventive activity most inventive persons are employed first, the research capability of a given number of such persons can be expected to increase as the total pool from which they are selected increases. This is the reason why L is an argument in the E function. Specification (10.1) is rather *ad hoc*, but we shall discuss the circumstances when it may be justified later in this chapter.

The optimum research intensity and the equilibrium innovation rate

Empirical evidence suggests that in the past two centuries or so the technology-producing sector has usually been expanding much faster than the conventional sector. It is instructive, however, to consider first the case of balanced growth. Accordingly, suppose that Y, K, and M all change at a common constant growth rate, to be denoted by g, and that L, N, and R change at another constant rate n. From (10.1) we have

$$g = \alpha + n \qquad (10.5)$$

where $\alpha = \dot{T}$, the innovation rate. Given the assumption of constant returns to scale, it follows from (10.2) that $\alpha = H(E/T, 1)$. Thus the rate α is constant, a requirement of balanced growth, only if T is proportional to E, say $T = \eta E$.

We also note that under balanced growth, gross investment in fixed capital equals $\dot{K} + \delta K + \dot{M} + \delta M$, or $(g + \delta)(K + M)$, where δ is the depreciation rate. Therefore the level of consumption is as follows:

$$C = F\{K, \eta E(M, L - N, L), N\} - (g + \delta)(K + M) \quad (10.6)$$

The level is at a maximum if the inputs K, M, and N are chosen to meet these first-order optimality conditions:

$$F_K = g + \delta \qquad (10.7)$$

$$\frac{M}{K} = \frac{1 - a}{a} \beta \qquad (10.8)$$

$$\frac{R}{N} = \mu \qquad (10.9)$$

Condition (10.7) gives the optimal intensity in the conventional sector, while (10.8) and (10.9) give the optimal sectoral distribution of the two resources capital and labour. Using these conditions we can find gross capital investment in each sector. We can also find the optimal (balanced-growth) research intensity, i.e. the expenditure on wages and investment in the technology sector as a proportion of the conventional output. Gross investment is, in the conventional sector,

$$(g + \delta)K = F_K K = \left(F_K \frac{K}{Y}\right) Y = aY$$

and, in the technology sector,

$$(g + \delta)M = (g + \delta) \frac{M}{K} K = (1 - a)\beta Y$$

Subtracting total investment from output gives consumption. Now, suppose that consumption is the same as the total wage income and that wage rates are the same in both sectors. Condition (10.9) enables us to find the wage income in each sector. Thus we have obtained both the investment and the wage element of the R&D expenditure. The research intensity i – the share of total output devoted to the technology sector – is in this case

$$i = (1 - a)\left\{\beta + \frac{\mu}{1 + \mu}(1 - \beta)\right\} \tag{10.10}$$

Since, in this model, a is the share of gross investment in the conventional sector, it can be expected to be less than 0.5. If presently observed values of M/K and R/N are any indication of their optimal values, then by (10.8) and (10.9) both μ and β are small and consequently both $1 + \mu$ and $1 - \beta$ are close to unity. The optimal research intensity can therefore be approximated as follows:

$$i \approx (1 - a)(\beta + \mu) \tag{10.10a}$$

Of central interest, however, is the magnitude of the equilibrium innovation rate α. We obtain it by recalling that $T = \eta E$, from which it follows that $\alpha = \dot{T} = \dot{E}$. From (10.3) we have in turn that $\dot{E} = \beta \dot{M} + (\mu + \gamma)n$. However, according to (10.5), $\dot{M} = \alpha + n$. Therefore $\alpha = \beta(\alpha + n) + (\mu + \gamma)n$, which gives

$$\alpha = \frac{\beta + \mu + \gamma}{1 - \beta} n = \alpha^* \tag{10.11}$$

The asterisk indicates that this is the equilibrium rate.

Two important implications of the result (10.11) can be noted immediately. One is that if β and μ are significantly less than unity, and there are indeed good grounds to place them between zero and 0.1, then the heterogeneity of the labour force with respect to inventive ability, represented by γ, may be a key factor determining the innovation rate. The other implication is that if the population of the TFA ceases to grow, so that $n = 0$, the rate α^* would also be zero. This particular implication may seem intuitively implausible. However, the same crucial relation between n and α^* also occurs in other models of innovation and growth, and therefore we shall consider it in more detail later in this chapter.

Human inventive and innovative heterogeneity and technological progress in the technology sector itself: two generalizations

Inventive ability is known to differ substantially between individuals. Figure 10.1 shows a possible distribution of the working population with respect to this ability ν. What matters for us is not the innate or natural inventive ability but the actual inventive ability, given possible environmental influences such as quality of schooling, family circumstances, and attitudes to learning, as well

173

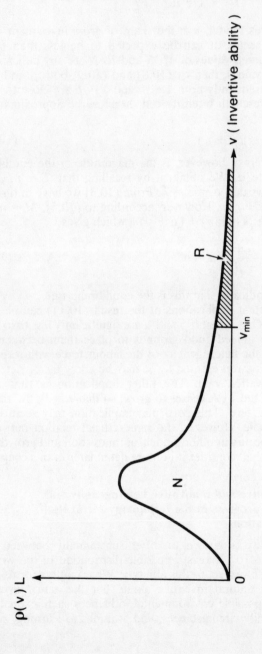

Figure 10.1 The distribution of the total working population *N* and *R* with respect to inventive ability *v*.

174

as the system of incentives and values encouraging potential inventors to make use of their potential. If the 'screening methods' of the 'appointments committees' are appropriate, the research workers would be represented by the shaded area in Figure 10.1 below the upper tail of the distribution $L\rho(v)$.

Our first modification of the Phelps model is to assume that this tail is a Pareto-type function, an assumption often adopted in economics (for example, to describe the upper tail of the distribution of income or wealth). In this case, $R/L = \text{prob}(v \geq v_{min}) = C_1(v_{min})^{-\lambda}$ where $\lambda > 1$ and C_1 is a positive constant. It follows that, for $v \geq v_{min}$, the underlying density distribution is $\rho(v) = \lambda C_1 v^{-\lambda-1}$. The total inventive ability of our R researchers can now be calculated:

$$V = L \int_{v_{min}}^{\infty} v\rho(v)\, dv = C_2 R^{1-1/\lambda} L^{1/\lambda} \tag{10.12}$$

where $C_2 = \{\lambda/(\lambda-1)\} C_1^{1/\lambda}$. This result justifies specification (10.3). Moreover, writing (10.3) as $E = C^\mu M^\beta V^v$, we have that $\mu = v(1 - 1/\lambda)$ and $\gamma = v/\lambda$. Consequently, in this case $\mu + \gamma$ in expression (10.11) for α^* would be equal to v; the innovation rate α^* would thus be independent of the ability variation parameter λ.

When the labour input in the R&D sector is expressed in units of ability-hours, as in (10.12), rather than in man-hours, as in the original Phelps model, then the E function need not be of the very restrictive Cobb–Douglas form to permit balanced growth. The least restrictive specification that would still be satisfactory is

$$E = E(M, TV) \tag{10.13}$$

where V is multiplied by T to account for technological change also enhancing the research capability of the researchers themselves. Equation (10.13) is our second modification of the model. Specifications (10.1) and (10.13) are now symmetric. Consequently, the elasticity of substitution between M and V, as between K and L in (10.1), need be neither unitary nor constant. Parameter β continues to be the elasticity of E with respect to M, and v now stands for the elasticity of E with respect to TV.

It is interesting to note the implication of replacing (10.3) by (10.13) for the magnitude of α^*, the key variable of the model. Let ε be the scale elasticity, assumed constant, of the E function. Hence

$$E = T^\varepsilon V^\varepsilon E(M/TV, 1) = T^\varepsilon V^\varepsilon e(m) \tag{10.14}$$

175

where $m = M/TV$. On a balanced growth path m is a constant and therefore $\dot{E} = \varepsilon(\dot{T} + \dot{V})$. However, $\dot{E} = \dot{T}$ and $\dot{V} = n$. Hence the balanced growth innovation rate would be

$$\alpha^* = \frac{\varepsilon}{1 - \varepsilon} \, n = \frac{\beta + v}{1 - \beta - v} \, n \qquad (10.15)$$

This result is similar to (10.11) where, incidentally, $\beta + \mu + \gamma$ is the same as ε in (10.15). In particular, despite allowing for (labour-saving) technological change in the R&D sector, the equilibrium innovation rate still remains proportional to the population growth rate. Conditions (10.7)–(10.9) for the optimal distribution of capital and labour between the two sectors, conventional production and inventive activity, also remain unchanged. However, for (10.15) to make economic sense, research activity must be subject to diminishing returns to scale ($\varepsilon < 1$).

Price's two laws and the 'technological revolution': the case of unbalanced growth

The term 'technological revolution' is one of those which are used often without being defined precisely. The term seems intuitively clear enough; it means a period of 'unusually' rapid innovation in a particular sector, country, or the world as a whole. Some authors distinguish major bursts of world innovative activity, such as that based on the steam engine and consequent mechanization, or electrical power and its applications, or inventions in electronics and telecommunication, or, recently, microelectronics and the use of robots. They refer to these bursts of innovations as technological revolutions – first, second, and so forth – in their own right. Such distinctions are sometimes useful to a social scientist by their virtue as indicators of the changing content of the innovation flow, with its implications for changes in the skills required, social stratification, social mobility, and the rate of spread of information and ideas. However, for the economist it is the rate of innovation flow as such, rather than the flow's specific content, which is of central interest.

What is meant, then, by an 'unusually' high rate of innovation? It must be a rate which cannot be sustained 'for ever', i.e. an innovation rate which is greater than the balanced-growth innovation rate, or α^* in (10.15). In terms of our two-sector economy, technological revolution can therefore be defined as a prolonged period of economic growth in which the technology sector is expanding faster than the conventional sector. Such unbalanced growth cannot be sustained for ever, but as long as it lasts it does give rise to $a > a^*$.

A convenient measure of the expansion of any sector of economic activity is a weighted sum of the growth rates of the inputs employed in that sector – labour and capital in the case of our present model. The data on the growth rates of these two inputs in the technology sector vary in quality among countries and between different periods. However, such records as we have indicate that, over the last two to three centuries, the world R&D sector has been expanding (i) nearly exponentially and (ii) much faster than the conventional sector. The empirical propositions (i) and (ii) are among the key stylized facts concerning technological change and long-term growth that have been (relatively) well established. According to the science historian, Derek de Solla Price:

> many numerical indicators of the various fields and aspects of science ... show with impressive consistency and regularity that if any sufficiently large segment of science is measured in any reasonable way, the normal mode of growth is exponential. (Price 1963: 4–5)

Price suggests that this steady exponential growth of the size of world science has been maintained for the past two to three centuries. Because of this long period of validity, he calls it the 'fundamental law of any analysis of science' (Price 1963: 5). We shall refer to it as Price's first empirical law. His second empirical law is as follows:

> depending on what one measures and how, the crude size of science in manpower or in publications tends to double within a period of 10 to 15 years. The 10-year period emerges from the catchall measures that do not distinguish low-grade work from high but adopts a basic, minimal definition of science; the 15-year period results when one is more selective, counting only on some more stringent definition of published scientific work and those who produce it. (Price 1963: 6)

The steady doubling every 10 to 15 years gives the growth rate of the world scientific membership as between 4.7 and 7.2 per cent per annum. Judging from the detailed country data for the past 50 years or so, the total labour input in both research and development has been increasing about as rapidly as the number of scientists alone. These data also indicate that the non-personnel

177

(real) expenditure in R&D has been expanding somewhat faster than the R&D personnel. These two growth rates in the period 1750–1975 have apparently been much higher than the growth rates of labour and capital in the conventional sector, which were roughly 0.7 per cent (the growth rate of the world population) and 1.7 per cent (the growth rate of the world GDP) respectively. It is this wide disparity in the sectoral growth rates, in favour of the technology sector, which above all underlines the phenomenon of 'technological revolution'. Such a disparity cannot be maintained for ever; in fact there has already been a significant slow-down in world R&D growth since about 1970. The balanced-growth solution of the previous section is relevant only with reference to an equilibrium configuration that was present in the distant past and will emerge in the (possibly less distant) future. However, the past two or three centuries represent a period of highly unbalanced growth that needs separate consideration. We shall do this in this section.

We retain the model specifications (10.1)–(10.4) of the previous section, except that (10.3) is replaced by (10.13). We also retain the assumption that the capital-to-output ratio in the conventional sector is constant. However, neither the savings ratio nor the growth rates of total output and its components need be constant. Let N_1 be the number of workers engaged in producing capital goods for the R&D sector. The assumption that the capital-to-output ratio is constant implies that, as in the previous section, the productivity of these N_1 workers is proportional to the aggregate level of technology T. Hence the growth rate of capital employed in the R&D sector is

$$\mathring{M} = \alpha + n_1 \tag{10.16}$$

The growth rate of the R&D labour input is

$$\mathring{R} = n_2 \tag{10.17}$$

Price's two empirical laws are (i) that n_1 and n_2 have both been significantly greater than n, and (ii) that they have been approximately constant.

Given n_1 and n_2, we can now obtain the innovation rate from equations (10.2) and (10.13).

Assume $H(E, T) = E^b T^{1-b}$ and $E(M, TV) = M^\beta (TV)^v$ where V is given by (10.12) and the elasticities b, β, and v are all constant. With these specifications the system determining α is as follows:

$$\mathring{\alpha} + b\alpha = b\mathring{E} \tag{10.18}$$

$$\overset{\circ}{E} = \beta \overset{\circ}{M} + v(a + \overset{\circ}{V}) \tag{10.19}$$

$$\overset{\circ}{V} = \frac{\lambda - 1}{\lambda} \overset{\circ}{R} + \frac{1}{\lambda} n \tag{10.20}$$

The growth rate of research effort is, in (10.18), the propelling force that determines the dynamics of the innovation rate. This growth rate is in turn determined by the growth rates of R&D inputs, the measure of the labour input taking due account of the variation in inventive ability, and the effect of innovation on efficiency in the research sector itself.

Superimposing the stylized facts (10.16) and (10.17) on the system (10.18)–(10.20) yields the following equation for $a(t)$:

$$(1 - \beta - v)a + \frac{1}{b} \overset{\circ}{a} = \beta n_1 + \left\{ \frac{1}{\lambda} n + \left(1 - \frac{1}{\lambda}\right) n_2 \right\} v \tag{10.21}$$

If the growth rates n_1 and n_2 are constant, then (10.21) can be solved explicitly to give

$$a(t) = a_{\text{TR}}^* + (a_0 - a_{\text{TR}}^*) \exp\{-b(1 - \beta - v)t\} \tag{10.22}$$

where $a_0 = a(0)$ and

$$a^* = \frac{\beta n_1 + v\{(1/\lambda)n + (1 - 1/\lambda)n_2\}}{1 - \beta - v} \tag{10.23}$$

From (10.22) it follows that $a(t)$ approaches a_{TR}^* with time. This a_{TR}^* stands for the 'equilibrium' component of the innovation rate in the course of the technological revolution.

A growth slow-down in world R&D activity has taken place during the 1970s and the 1980s. If the slow-down means that the optimal ratios of R/N and M/K are about to be or have already been achieved in the TFA, we can use the optimality conditions (10.8) and (10.9) for estimating the values of the parameters that appear in (10.23). These conditions are that $v(1 - 1/\lambda) = R/N$ and $\beta(1/a - 1) = M/K$. Guided by empirical evidence we also assume that $R/N = M/K$ and $n_1 = n_2$. Consequently, the following relationship can be obtained:

$$a_{\text{TR}}^* = \frac{(\lambda - 1)n_1 + (1 - a)n}{(\lambda - 1)(1 - a)(N/R) - (\lambda - 1)a - \lambda(1 - a)} \tag{10.24}$$

179

It should be noted that the higher is the value of λ the lower would be the proportion of highly innovative individuals; the case of $\lambda = \infty$ is the limiting situation where no very innovative talent is present (Figure 10.2).

According to Price, the rate n_1 has been somewhere in the range 4–8 per cent. Given this range, it is instructive to find the values of λ for which the rate α_{TR}^*, as given by (10.24), would be approximately equal to the innovation rate actually observed. These values are presented in Table 10.1. Since our knowledge of the optimal ratio R/N is uncertain, the table provides the values of λ for a range of R/N from 1 to 5 per cent.

A consistency test of the model and Lotka's law

The instructive point of this numerical example is the result that, for the probable values of α_{TR} and R/N, the model we discuss predicts λ to be only somewhat greater than unity. If there was independent evidence indicating that the λ actually observed is in fact not far from unity, the model would pass an important empirical test.

The trend growth rate of GDP per man-hour, which can be taken as a measure of α, was 2.3 per cent per annum in the United States in the period 1870–1970 (Maddison 1979). Since the growth rate was fairly stable during that period, it can also be taken as a measure of α_{TR}^*.

An indication of the value of λ is provided by studies of the frequency distribution of scientific productivity. A pioneer investigation of this type was made by Lotka (1926). The result of his investigation, later repeated and confirmed by several others, is the finding that the number of scientists producing m papers within their lifetime is approximately proportional to $1/m^2$. The number of publications or the number of inventions is, of course, only one of several possible measures of inventive power. The measure may be a poor guide for judging the weight of the contribution to science or technology of any particular individual, but a good guide for the 'representative' scientist (inventor).

If we take m as a measure of the inventive ability, denoted by v in equation (10.12), Lotka's law asserts that our frequency distribution $\rho(v)$, the distribution specified in (10.12) as $C_1 v^{-\lambda-1}$, is proportional to v^{-2}, implying that $\lambda = 1$. Several other investigators have since repeated such publication counts. According to Price, they all confirm Lotka's result, 'which does not seem to depend upon the type of science or the date of the index volume' (Price 1963: 43). Moreover, Lotka's law is known to overestimate

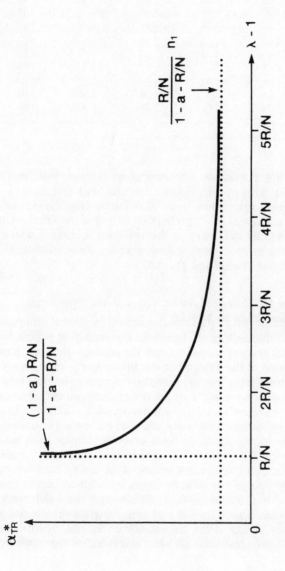

Figure 10.2 The relationship between the innovation rate α_{TR} and the ability variation parameter λ, as given by equation (10.24).

Table 10.1 The values of λ as implied by (10.24), given α^*_{TR}, n_1, and R/N, and assuming that $a = 2$ and $n = 0.7$ per cent

α^*_{TR} (%)	n_1 (%)	λ				
		$R/N = 0.01$	$R/N = 0.02$	$R/N = 0.03$	$R/N = 0.04$	$R/N = 0.05$
1	4	1.02	1.04	1.06	1.09	1.13
	6	1.02	1.04	1.07	1.11	1.15
	8	1.02	1.04	1.07	1.12	1.19
2	4	1.01	1.03	1.05	1.06	1.08
	6	1.01	1.03	1.05	1.07	1.08
	8	1.01	1.03	1.05	1.07	1.10
3	4	1.01	1.02	1.04	1.05	1.07
	6	1.01	1.02	1.04	1.06	1.08
	8	1.01	1.02	1.04	1.06	1.09

Note: The values of λ are rounded to two decimal places.

somewhat the proportion of researchers with a high m (Price 1963: 46–9). This in turn implies that the 'true empirical' λ is in fact somewhat greater than unity. It is interesting, indeed remarkable, that such values of λ also happen to be the requirement of the theoretical model discussed in the previous section. Lotka's law thus seems to provide empirical support for this particular theory of technological change and growth.

The Second Hat-Shape Relationship and the hypothesis of innovation limits to growth

The story of technological change and growth that is told by the theory of this chapter is one in which the technological revolution is a phenomenon of the TFA when the key resource ratios R/N and M/K are rising fairly fast to reach their optimum levels, and when the key growth rates n_1, n_2 and α are temporarily significantly higher than their balanced growth magnitudes n and α^*. The last two centuries are not the only ones when inventive activity, in terms of the inputs used, has been expanding faster than conventional activity. The history of science and technology provides ample evidence of significant bursts of inventive and innovative work in Europe in the Middle Ages, as well as in the ancient civilizations of the Middle East, China, and the Mediterranean. However, what makes the present technological revolution qualitatively quite unique is the circumstance that the growth rates n, n_1, and n_2 have apparently all been much higher than ever before over a prolonged period, giving rise to a correspondingly much higher innovation rate with profound implications for the pace of economic and social change in much of the world.

In the past century or two the relative size of the science and technology sector has been rising rapidly, but this change of size apparently did not influence the innovation rate very much, which remained fairly stable in the TFA. This stylized empirical fact agrees well with our equations (10.22) and (10.23), provided that the growth rates of research inputs, rather than their levels, influence the innovation rate. In equation (10.23), these levels could influence α only through the parameters β and v. Therefore we can deduce that these parameters have been almost independent of the ratios M/K and R/N respectively and that the actual α was near to α_{TR}^*.

Given the apparent stability of β and v so far, it is fair to assume that the two parameters will remain in future about the same as they were in the past. However, the future will bring about two important new phenomena: (i) an inevitable fall in the employment growth rates n_1 and n_2 to about n, as the technology sector ceases to claim an increasing share of resources, and (ii) an equally inevitable fall of the rate n itself to about zero as the world population growth comes to an end. It is interesting, in the light of our theory, to find what the impact of these two phenomena on the innovation rate in the TFA will be.

Case (i): end of the faster growth of the technology sector

From (10.23) it follows that for $n_1 = n_2 = n$ the target innovation rate would be

$$\alpha^* = \frac{\beta + v}{1 - \beta - v} n \qquad (10.25)$$

According to (10.22) the actual rate $\alpha(t)$ would be falling from α_{TR}^* to α^* exponentially at a speed depending on the value of $(1 - \beta - v)b$. Let us express α^* in terms of α_{TR}^*:

$$\alpha^* = \frac{(\beta + v)n}{n_1 + v\{(1/\lambda)n + (1 - 1/\lambda)n_2\}} \alpha_{TR} \qquad (10.26)$$

To illustrate the possible size of the fall, suppose that on a balanced growth path $M/K = R/N = $ constant. Suppose also that in the past $n_1 = n_2$, $n = 0.7$ per cent, and $a = 0.2$. On using the

183

optimality conditions, namely that $\beta = \{a/(1-a)\}M/K$ and $v = \{\lambda/(\lambda-1)\}R/N$, expression (10.26) implies that

$$\alpha^* = \frac{\lambda + \{(\lambda-1)/(1-a)\} an}{1 + \{(\lambda-1)/(1-a)\}n_1/n} \alpha_{TR}^* \qquad (10.27)$$

Lotka's law suggests that the value of λ is not much greater than unity; suppose that it is at most 1.2. The ratio n_1/n is unlikely to have been greater than 10. Substituting these two numbers into (10.27) gives the lower limit for α^*, which turns out to be about $0.35\alpha_{TR}^*$. Thus the upper limit for the innovation slow-down is from α_{TR}^* down to $0.35\alpha_{TR}^*$. The size of the slow-down is sensitive to the value of λ. Taking $\lambda = 1.1$ and retaining the values of the other parameters gives $\alpha^* = 0.49\alpha_{TR}^*$.

Apart from the magnitude of the fall of α from α_{TR}^* to α^*, it is also of interest to inquire about its speed. We note from (10.22), where α_{TR}^* should be substituted for the initial rate α_0 and α^* for the target rate α_{TR}^*, that

$$\alpha(t) - \alpha^* = (\alpha_{TR}^* - \alpha^*) \exp\{-(1 - \beta - v)bt\} \qquad (10.28)$$

The values of $1 - \beta - v$ and b are between zero and 1. If the ratios R/N and M/K were to remain in future at about their levels observed in the TFA at present, then we can deduce from the optimality conditions (10.9) and (10.10) and Lotka's law that $0 < \beta \ll v < 1$.

The speed of innovation slow-down, following the end of a faster growth of the technology sector, can be measured by the difference $\alpha(t) - \alpha^*$, expressed as a fraction of $\alpha_{TR}^* - \alpha^*$, that will still remain after a period Δt years. It is clear from Table 10.2 that for $(1 - \beta - v)b \geqslant 0.1$, the fall from α_{TR}^* to α^* would be fast. However, the value of the research elasticity b is not known with any accuracy. It may well be low enough to result in a slow fall. A low value of b occurs when the innovation output is

Table 10.2 The ratio $(\alpha - \alpha^*)/(\alpha_{TR} - \alpha)$ after Δt years from the start of the innovation slow-down

	$(\alpha - \alpha^*)/(\alpha_{TR} - \alpha)$		
$(1 - \beta - v)b$	$\Delta t = 10$	$\Delta t = 20$	$\Delta t = 50$
0.10	0.37	0.13	0.007
0.25	0.10	0.007	0.000

influenced strongly by the research accumulated in the past and is not particularly responsive to changes in the current amount of research, even allowing for a suitable time-lag.

Case (ii): end of the world population growth

Should the population of the TFA cease to grow, the model predicts that α^* would then be zero. The innovation rate would of course be positive at any finite time, but it would be falling continuously from α^*_{TR} to near zero after a sufficiently long period. This is clearly an interesting prediction, especially since an end to population growth in the TFA, and eventually in the whole world, may not be far off. It is therefore important to discuss the model's implications for the innovation rate and economic growth in this case.

If the population is constant and a constant proportion of it is engaged in R&D, our variable V, denoting the total number of researchers corrected for their inventive ability, is also constant. The amount of research is nevertheless increasing. This is because technological progress still takes place, increasing not only the output of the conventional sector, and therefore investment M, but also the productivity of inputs M and V in the research sector. Both M and TV are in fact rising at a common rate equal to α. However, the scale elasticity $\beta + v$ of the E function must be less than unity if equation (10.23) is to make economic sense. Otherwise not just the levels of technology and GDP but their growth rates as well would be rising to infinity, something that the empirical evidence seems definitely to rule out. However, if $\beta + v < 1$, the research output would grow at a rate less than α. The ratio E/T would therefore be declining. Hence $\alpha = H(E/T)$ would also be declining.

It should be noted that absolute annual additions to the technology level, if following the rule that $\Delta T = H(E,T)$, would continue to be increasing with time. However, these additions would be increasing at a falling percentage rate, to become eventually nearly constant. Total conventional output, capital stock, and consumption would all also continue to increase. However, instead of increasing at a nearly geometric rate, as they did when the technology sector itself had been expanding nearly exponentially, they would be increasing at a nearly arithmetic rate. The innovation limits to growth should therefore be understood to mean a very long-term growth slow-down, both technological and economic, but not necessarily an end to economic growth. The model does not imply that there is any finite upper limit to the level of technology or conventional output; such a limit is known to arise if some essential inputs were both non-reproducible and difficult to substitute for by reproducible inputs, with the relevant elasticities of substitution being less than unity. In our model the labour input is the only natural resource, but it is one which is reproducing itself; it is not non-reproducible.

It seems obvious, or at least possible, that if the world is finite, everything is finite, including the scientific and technological knowledge that is still to be discovered. In the model the innovation limits of this ultimate kind are implicitly assumed to be so distant as to have no impact on the inventive productivity of our researchers. However, this factor may be expected to reinforce the innovation slow-down in due course.

The Second Hat-Shape Relationship

In Chapter 9 we discussed the variation of innovation rates among countries at different levels of development in any relatively short period of time, such as a year or a decade. We noted that such cross-country variation forms a hat-shaped pattern, with the medium-developed countries tending to experience faster innovation than both the least and the most developed countries.

Our discussion in this chapter is limited to the TFA of the world; the central question is how the area's innovation rate changes over time in the course of centuries. This is thus 'one-country' dynamic analysis. This analysis indicates that the pattern of change of the innovation rate over time may also be hat-shaped. However, the First Hat-Shape Relationship is an empirical law that is given a theoretical interpretation. The Second Hat-Shape Relationship as seen in Figure 10.3 is an implication of a particular model of innovation and growth. Its acceleration and steady growth segments correspond well to the observed reality. However, its slow-down part is yet to be tested.

The five periods or stages distinguished in Figure 10.3 have the following characteristics (Gomulka 1971).

(i) The growth rates n_1, n_2, and n are all low: consequently the innovation rate is also low.

(ii) n_1, $n_2 \gg n$, and the population growth rate n is high: the actual innovation rate increases from the low level of the previous period, approaching the target innovation rate α_{TR}^* at the end of period (ii).

(iii) n_1, $n_2 \gg n$ and $\alpha = \alpha_{TR}^*$: this is a period of balanced growth in the sense that all the growth rates are fairly stable,

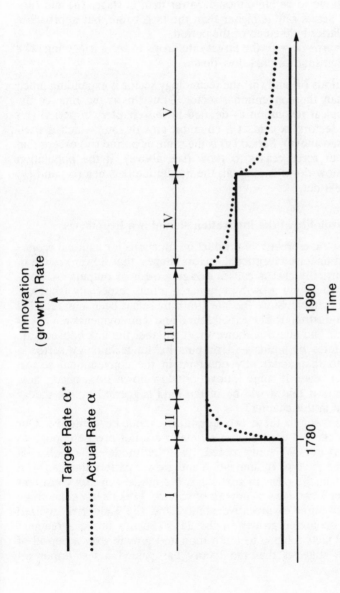

Figure 10.3 The target and actual innovation rates over time in the TFA. The dates and magnitudes are chosen for illustrative purposes.

187

but unbalanced in the sense that the technology sector (more generally, the sector upgrading the quality of the labour and producible material inputs) is expanding much faster than the conventional sector.

(iv) $n_1 = n_2 = n$, but n is still high: the target innovation rate may continue to be high, though lower than in stages (ii) and (iii). The actual rate is higher than the target rate, but approaches the latter by the end of the period.

(v) $n_1 = n_2 = n = 0$: the target rate drops to zero, triggering off a further innovational slow-down.

In periods (ii) and (iii) the technology sector is expanding much faster than the conventional sector. Thus this is the time of the technological revolution as defined in this chapter. In period (iv) the two sectors expand at a common growth rate, which is itself falling (see above). Period (v) is the same as period (iv) except that the labour force ceases to grow (see above). If the population growth slow-down is gradual, the borderline between (iv) and (v) is not clear cut.

The plausibility of the innovation slow-down hypothesis

Econometric estimates of production functions for national economies and major conventional sectors suggest that slower growth of inputs virtually always causes slower growth of outputs. A slow-down in the trend growth of research inputs, especially labour, is clearly inevitable. Since the late 1960s there has been a marked fall in the industrial R&D expenditure and employment in OECD countries. The question, however, is whether the link between the growth rates of inputs and outputs in the technology sector is similar to that apparently observed in the conventional sector. Moreover, even if some innovation slow-down does occur, how plausible is it that it will be of the kind suggested by the theory discussed in this chapter?

The answer to these two questions cannot be definitive. Our growth data and econometric estimates are not precise enough to be otherwise. We only noted that the model is capable of generating the type of innovation and growth pattern that has been observed in the past. In particular, the model appears consistent with Price's two laws of growth of science, Lotka's law concerning the distribution of inventive ability, and the Kaldorian stylized facts of economic growth in the TFA. Such a broad agreement with key facts relating to innovation and growth over a period of centuries suggests that the theory has passed a test important enough to be taken seriously both as a plausible interpretation of economic growth in the past and as a useful vehicle for making predictions about the growth in the future.

The crucial assumptions of the theory are those which underlie the technology production function – the function which links research inputs and the innovation rate. Following Phelps (1966), we have assumed that $T = H(E, T)$. Suppose that this H function is specified as follows:

$$\dot{T} = E^{b(t)} T^{1-b(t)} \qquad 0 < b(t) < 1, \; t = T/E \qquad (10.29)$$

Consider now the plausibility of the innovation slow-down hypothesis when the amount of research effort E is a given constant. In this case the ratio T/E increases. The proportional rate $\dot{T}/T$ would then fall if the elasticity $b(t)$ is either constant or rising. However, $\dot{T}/T$ need not fall if $b(t)$ is declining. The elasticity $b(t)$ is declining if new additions to the technology stock are increasingly less dependent on current research and more dependent on accumulated past research. The evidence is again not definitive, but if anything it seems to suggest an opposite trend – an increasing weight of the current or immediately past research in innovation (see Chapter 3). If this were the case, the slow-down hypothesis would be given an added credibility. Be that as it may, the hypothesis clearly rests on the assumption that the weight of E in the production of new innovations is not falling (or not falling fast).

Another necessary condition for the innovation limits to growth hypothesis is that the weight V of the human effort E in producing research is not falling fast. If that weight, represented by the elasticity v, were to fall continually, the growth-constraining influence of the constant number of researchers would be continually reduced. If the scale elasticity in the R&D sector is assumed to be constant, the influence of machines in the research sector would then be increasing. The elasticity v falls when the labour–capital substitution elasticity is greater than unity, i.e. when machines can 'easily' substitute for researchers. Little is known about the value of this elasticity, however.

The technology function used by Mansfield (1968), and more recently by Dasgupta and Stiglitz (1980), assumes that

$$\dot{T} = A^a$$

$$A = \int_{-\infty}^{t} \exp\{-\delta(t - \tau)\} E(\tau) \, d\tau \qquad (10.30)$$

where $E(\tau)$ is the R&D expenditure at time τ, the expression

189

$\exp\{-\delta t - \tau)\}E(\tau)$ is the expenditure's effective value at time t, taking into account depreciation of any past research at a rate δ, and A is the effective R&D expenditure accumulated until time t. Both a and δ are assumed to be constant. With this specification there would be no loss from concentrating a given amount of total research expenditure in a short interval of time, and no gain from a smooth research effort. However, despite these differences the specification (10.30) also implies the presence of an innovation slow-down if $a < 1$, the restriction necessary to allow balanced growth. To see this implication note that, from the definition of A, $\dot{A} = E/A - \delta$. Assuming research expenditure to be a constant proportion of conventional output, the latter to be related to T by equation (10.1), and the capital-to-output ratio to be constant, we have that $\dot{E} = \dot{T} + n$. Moreover, $\dot{T} = a\dot{A}$. The economy described by these three equations tends to approach a balanced-growth path on which $\dot{A} = \dot{E}$. Hence, on that path $\dot{T} = a(\dot{T} + n)$ and

$$\mathring{T} = \frac{a}{1 - a} n \tag{10.31}$$

This result is qualitatively the same as that obtained with both Phelps's original technology function and its modified version developed in this chapter. Thus this model also implies that once n is zero the innovation rate, while always positive, would continually be falling to zero. For this to happen the elasticity a must have an upper bound less than unity and it must not move 'too quickly' towards that upper bound. (If a increases fast enough the slow-down is only postponed until the magnitude of a is near the upper bound.)

The widespread and increasing application of computers and robots has given rise to the view that massive increases in labour productivity can be expected in those sectors of economic activity that used to be highly labour intensive, such as banking, education, trade, and public administration. Many jobs in the so-called traditional industries, producing goods which have low income elasticities of demand, may also disappear. The information-processing industry has already become or is expected to become the dominant employer, perhaps by the end of the twenty-first century relegating manufacturing, in terms of employment, to more or less the status of agriculture by the end of the twentieth century. Such structural changes may well occur. However, some writers also predict or suspect that a significant upturn in the rate of growth of labour productivity will take place for the TFA as a whole, not just in those selected activities that are particularly suitable for automation and computerization. The hypothesis of innovation limits to growth flies very much against such a prediction.

In the period 1973–80 labour productivity growth has slowed down considerably in Western Europe (except the United Kingdom) and Japan, as well as in the USSR and Eastern Europe, very much in agreement with the catching-up hypothesis. However, the reasons for some slow-down in the United States are less clear. The period has been marked by particularly rapid changes in relative prices and demands, forcing relatively frequent adjustments in production techniques and outputs. Some adjustments are costly, thereby depressing factor productivities, labour productivity among them. Having said that, there is so far no evidence to contradict the innovation limits to growth hypothesis (Gomulka 1971, 1988). Computers and robots displace some jobs as steam or electric motors used to do in the past, but they also create jobs for workers able to design, produce, and service them. Neither is it clear what the overall productivity effect of these conflicting employment implications of new technology will be in the years to come. It is possible that it will be as the innovation limits hypothesis predicts.

Conclusion

Kondratieff's waves or cycles refer to possible bursts of innovation activity and growth some 50–60 years apart. According to the theory presented in this chapter, such waves, if they exist, would be only relatively small fluctuations on the surface of what may be called a superwave of the world's innovation activity. To put it differently, the 'technological revolution' of the past two centuries is suggested to be, by historical standards, just a brief transistory phenomenon, and the innovation rate is expected to decline in the course of the next century or two to the levels observed in 'normal' pre-revolutionary times.

Chapter eleven

Evidence and microeconomics of the international technology transfer

International distribution of R&D activity and new patents

Given the very unequal distribution of accumulated material wealth and human skills among nations, it is to be expected that most R&D and inventive activity of the world would be located in the technology frontier area (TFA) and near to it. This is indeed the distribution pattern which emerges from the data. The share of R&D expenditure in GNP in developing countries is only about a fifth of the respective share in the remaining countries, indicating that the concentration of world R&D activity in developed countries is much greater than that of world income. Taking the world total as 100 per cent, developing countries accounted in the 1980s for about 70 per cent of the population, 10 per cent of the R&D personnel, and 4 per cent of the R&D expenditure.

A finer classification of countries than that between developing and developed was proposed by Robert Evenson (1984). His purpose was to identify the distribution of various types of new invention patents and trademarks among (a) developed market economies, (b) planned economies (including Yugoslavia but excluding China, Vietnam, Mongolia, and North Korea), (c) medium-developed market economies, and (d) developing economies. Category (c) includes the newly industrialized countries of the Far East and Latin America, as well as Spain, Portugal, Greece, and Israel. Tables 11.1 and 11.2 contain a selection of the data compiled by Evenson that are of particular interest here. The number of invention patents granted to nationals in countries in categories (c) and (d) was quite marginal in 1967 and has declined further, both in absolute terms and as a percentage of the world total, in the 1970s and the 1980s. That number has substantially increased in the USSR, lifting the share of planned economies from 26 per cent in 1967 to 42 per cent in the years 1976 and 1980.

Table 11.1   Invention patents granted to nationals at home

	1967		(1976 + 1980)/2	
	Number (thousand)	Share (%)	Number (thousand)	Share (%)
Developed market	117.5	68.8	112.7	55.1
Planned	44.7	26.2	85.1	41.6
Medium developed	8.0	4.7	6.3	3.1
Developing	0.6	0.4	0.3	0.2
Total	170.8	100.0	204.7	100.0
USA	51.3	30.0	40.7	19.9
FRG	5.1	3.0	9.8	4.8
Japan	13.9	8.1	35.2	17.2
UK	9.8	5.7	7.0	3.4
France	15.2	8.9	8.4	4.1
Switzerland	5.4	3.2	2.5	1.2
Netherlands	0.3	0.2	0.4	0.2
Italy	9.1	5.3	1.8	0.9
USSR	24.0	14.0	67.0	32.7

Source: Compiled by the author on the basis of the data reported by Evenson (1984)

Table 11.2   Invention patents granted to nationals abroad by country of origin

	1967		(1976 + 1980)/2	
	Number (thousand)	Share (%)	Number (thousand)	Share (%)
Developed market	181.2	95.4	192.3	95.6
Planned	6.2	3.2	6.4	3.2
Medium developed	1.8	0.9	1.7	0.8
Developing	0.8	0.4	0.8	0.4
Total	190.0	100.0	201.2	100.0
USA	74.0	38.9	72.3	35.9
UK	17.6	9.3	12.6	6.3
FRG	41.8	22.0	35.5	17.6
Japan	6.8	3.6	20.5	10.2
France	14.4	7.6	12.6	6.3
Switzerland	12.5	6.6	10.4	5.2
Netherlands	7.3	3.8	5.9	2.9
Italy	5.6	2.9	5.6	2.8

Source: Compiled by the author on the basis of the data reported by Evenson (1984)

However, the usefulness of data on patents granted to nationals suffers from considerable differences in granting standards between countries and over time, as well as from changes in the composition of new patents in terms of their potential economic significance. The percentage of inventions for which no patents are sought also varies between countries and over time. Most such inventions in planned and developing economies are only or mainly of potential domestic use and would not be given patent rights in the most developed countries. The category of internationally recognized inventions is therefore of particular significance, as these inventions are usually those capable of moving the world's technology frontier. A proxy for inventions of this category is provided by the invention patents granted to nationals abroad. Table 11.2 gives the relevant data. Some of these inventions would be patented in many countries. This practice of multiple patenting can be considerd to bias upwards the inventive output of countries which patent more aggressively, or it can be viewed as a way of giving a greater weight to more significant inventions. Be that as it may, in terms of this proxy the contribution of the few developed market economies has been and is overwhelmingly dominant, standing at about 95 per cent of the world total in the years 1960–80, and probably remaining at or near that level for much of the twentieth century.

Among the developed countries, the United States is the largest contributor, accounting for 40 per cent of the world total number of invention patents granted to nationals abroad in the post-war period. In the late 1980s and 1990s that share is likely to be between 20 and 30 per cent. The Japanese share, while increasing rapidly, remained significantly below that of the FRG until the 1980s. The productivity data for 1978, in terms of patent applications abroad per scientist and engineer, are as follows (United States = 100): 341 for the FRG, 314 for the United Kingdom, 216 for France, 159 for all developed countries except the United States, 38 for Japan, and 5 for the planned economies (Slama 1983: Table 8). The practice of aggressive multiple patenting by Western European countries may account for much of their higher invention productivity as expressed in terms of this particular productivity measure. However, until recently Japanese productivity has been so unusually low among the developed countries that it must in part reflect the strategy of devoting large R&D resources to learning and imitating rather than original invention. Since the 1980s that strategy has been changed, as Japan can gain more from original invention which the country is now in a position to undertake.

The data discussed above give rise to the following three generalizations.

(i) Almost all world original and economically significant inventive activity takes place in the developed market economies of Western Europe, North America, and Japan.

(ii) In the countries outside the TFA, the share of R&D expenditure in GNP declines as the size of the technological gap between a particular country and the TFA increases. These countries are thus even less inventive when they are poor. Moreover, their limited R&D resources are mainly engaged in inventing of the adaptive and imitating type, and this tendency is more pronounced the larger the gap.

(iii) Countries with centrally planned economies – especially the USSR, the German Democratic Republic (GDR), and Czechoslovakia – are something of an exception to rule (ii). Their R&D effort and extent of industrialization are already similar to those of the developed market economies, but their contribution to the flow of internationally significant inventive output continues to be disproportionately small.

The technologically dominant position of the United States and Western Europe is in part sustained by the research talent originating from all other countries, which is attracted to the most developed countries by better quality research facilities in addition to higher salaries. From the point of view of maximizing the world innovation rate, this two-way pattern of brain drain on the one hand and technology transfer on the other must clearly be an economically efficient arrangement.

For most of the twentieth century the combined inventive output of Western Europe, as measured in terms of patents, was about the same as or greater than that of the United States. It is therefore interesting that in the 1880s the output per man-hour in US manufacturing began, nevertheless, to exceed that in UK manufacturing, the highest in Western Europe at that time, to reach a level of some twice the UK level by the late 1930s. In the meantime labour productivity in other Western European countries reached the UK level, implying that, in the years 1930–60, the United States had a very significant productivity advantage over the whole of Western Europe. There were many reasons for these developments – the asymmetrical effects of the two World Wars are the most obvious as well as perhaps the most important. Another possible reason was that the large-scale technologies were adopted faster in the United States, where state markets were much better integrated than in the highly divided Western Europe.

195

Political rivalries forced the major European countries to build parallel defence-related industries while tariffs and cultural differences helped to sustain parallel consumer industries. Since the 1950s these factors have gradually begun to abate, giving way to rapid growth of intra-industry intra-European trade, which in turn induced powerful changes in the European market structure that must have been conducive to the adoption of lower-cost large-scale technologies. Since 1945 the productivity gap between continental Western Europe and the United States has been reduced in part as a result of these changes as well as through the standard channels of international technology transfer.

Channels, costs, benefits, and the role of local R&D activity

Given that most new technology is produced in or near the TFA (which constitutes at present a relatively small part of the world in terms of population or territory), it is important to understand how that technology is diffused worldwide. We began discussing this topic in Chapter 9 where the First Hat-Shape Relationship was introduced. This relationship is an empirical law which indicates that international technology transfer is the dominant source of innovation and growth in most countries behind the TFA. In particular, it was noted that the medium developed countries are those which record the highest gains in productivity growth. They are thus capable of catching-up with the TFA quickly, even though their own inventive activity is small. The innovation dynamics during this technological catching-up motivates our interest in the economics of international technology transfer. Technology transfer is also important for the highly inventive developed market economies. A large or medium economy of this category – such as Japan, the FRG, France, or the United Kingdom – produces or is capable of producing almost all conventional goods for which there is significant home demand, but, on its own, accounts for a relatively small fraction, usually less than 10 per cent, of the world total output of inventions. Access to foreign inventions is therefore also necessary for these economies.

There are many ways in which an invention can be transferred and adopted across countries. The following two-way classifications are now sometimes used: disembodied versus embodied technology transfer (Gomulka 1971), and commercial versus non-commercial transfer (Hanson 1981). Disembodied type is in the form of a verbal description, sketch, research paper, blueprint, or a book describing a new product which has already been made or machinery and equipment that incorporates a new production process. Sometimes a technology transfer involves a piece of hardware, a blueprint describing how the hardware works, and a manual giving operating instructions; it is then both embodied and disembodied. Access to data on imports of machinery and equipment from developed countries has led empirical studies of the size and economic implications of technology transfer to focus on its embodied form. Transfer in this form is necessarily the dominant channel for a country which has an insignificant R&D and engineering capability of its own and which therefore has to import both the know-how and the hardware. The R&D and engineering capability also defines the capacity to absorb new technology in its disembodied form. Most medium and highly developed countries therefore have a large absorptive capacity of this category. These countries would typically have the skilled personnel and sophisticated equipment necessary to make the machinery needed to produce a new foreign invented product or use a new foreign-invented production process. On these a priori grounds, disembodied diffusion to these countries can be expected to be an important, even dominant, form of their technology imports.

Commercial transfer is subject to trade contracts between the parties concerned: individuals, companies, or governments. The value of such contracts would be recorded in a country's trade statistics. However, typically only the inventive output of the last several years is covered by patents or is otherwise protected. Most older technology and almost all (new and old) science is (almost) freely available and would therefore not be subject to commercial contracts.

Whether commercial or not, technology transfer almost always involves resource costs. These are all the costs that must be incurred to obtain a foreign technology and to ensure its effective utilization. Typically, they comprise royalty fees and the cost of transmitting and absorbing all the relevant know-how, which includes the peripheral information relating to plant design and construction, equipment installation, organization and operation of production, quality control, modifying the technology, and training labour.

A detailed discussion of these costs is given by Teece (1977) among others. In his case study of twenty-six investment projects involving international technology transfer, the total transfer costs were found to average 19 per cent of total project costs, with the percentage ranging from 2 to 59 per cent. Teece identified four variables as particularly important in influencing the size of the transfer costs.

(i) the extent to which the technology is understood by the transferee;
(ii) the age of the technology;
(iii) the number of firms utilizing the same or similar technology;
(iv) the cultural and systematic characteristics of the importing country.

His econometric estimates imply that 'transfer costs decline as the number of firms with identical or "similar and competitive" technology increases, and as the experience of the transferee increases' (Teece 1977: 253).

Let T_h represent the technology level at home, T_F the level in the TFA, and T_m the level of technology imported. Teece's econometric estimates suggest that the international transfer costs rise with T_m, given T_h and T_F. Domestic technology transfer costs in his sample were also considerable, although in most cases they were significantly lower than the international transfer costs. The sample variation of transfer costs, both domestic and international, was found to be very wide. However, Teece noted that 'the most difficult and hence costly technology to transfer is characterised by very few previous applications, a short elapsed time since development, and limited diffusion' (Teece 1977: 249). This would imply that the costs T_m of transferring technology are particularly high when $T_F - T_m$ is low, as T_m is then near to, in Teece's terminology, the 'leading edge' technology, which is as yet relatively untried and in a state of flux. Also, royalty fees for such technology can be expected to be higher, as it would be in limited supply and may (be thought to) offer potentially larger benefits.

Transfer costs can be viewed as a form of investment to buy or otherwise obtain a benefit-producing asset. Given total investment resources, technology transfer projects could compete with other investment projects and those promising sufficiently high rates of return would be accepted. A macromodel to be discussed in the following chapter attempts to find an optimal distribution of investment resources between conventional production, domestic R&D, and importing foreign technology, taking relevant costs and benefits into account. The benefits associated specifically with technology transfer will be identified there. The costs and benefits of technology transfer are also apparent in the micromodels to be discussed below. (Technology transfer costs are discussed further in the next chapter.)

The Krugman model and its extensions

It is natural to expect the world's first production of a new product to be set up in the country in which that product was invented. This is indeed usually the case. The exceptions are the cases when the product is fairly sophisticated but is invented in a country which does not have the skilled labour or equipment necessary to produce it. Nelson and Norman (1977) developed a model of international product cycle based on the point that important new products and processes are also rarely in a state of the art ready for widespread use. In the first phase of the cycle a great deal of R&D work is still required, often involving the need to invent new tools and components in the supplying industries, before the inherent superiority of the new product or process becomes fully apparent. The size of the skilled R&D and supervision personnel at that stage is still large. Moving it to less developed countries, in order to take advantage of the low wages of the less skilled production workers there, may be unprofitable in that phase. Only when production becomes standardized and, as a consequence, less skill intensive, may setting up production facilities in a low-wage country prove comparatively more profitable. There is also the point that novelty prices tend to be high for what the new products offer, and therefore the markets for these products are initially both small and located almost exclusively in the rich developed countries.

An extremely simplistic yet quite instructive model of product innovation and international diffusion was developed by Paul Krugman (1979). In this model the world is assumed to consist of an innovating North and an imitating South. The innovations are in the form of new products and methods of making these products only; the ways of making all existing goods do not change. Any new products are trade diffused from North to South as soon as they are invented, but their production methods are transferred with a time-lag which is assumed to be common for all methods. Both the inventive activity in the North and the North–South technology transfer activity are costless. The rate of innovations in the North is measured by the percentage of all products which new products represent. This rate is assumed to be a given constant. Homogeneous labour is the single input. Furthermore all industries are perfectly competitive. Workers worldwide not only have identical skills but their preferences as consumers of the invented goods are also the same. Given these assumptions, the equations of the model are as follows:

$$p_i = w_N \qquad i = 1, 2, \ldots, n_N \tag{11.1}$$

199

$$p_i = w_S \qquad i = n_{N+1}, \ldots, n \tag{11.2}$$

$$\dot{n}_N = \alpha n - \dot{n}_S \tag{11.3}$$

$$\dot{n}_S = \beta n_N \tag{11.4}$$

$$n = n_S + n_N \tag{11.5}$$

$$U = \left(\sum_{i=1}^{n} c_{ij}^{\theta} \right)^{1/\theta} = \max ! \qquad 0 < \theta < 1 \tag{11.6}$$

subject to

$$w_j = \sum_{i=1}^{n} p_i c_{ij} \qquad j = N, S \tag{11.7}$$

$$L_S \sum_{i=1}^{n_N} p_i c_{iS} = L_N \sum_{i=n_N+1}^{n} p_i c_{iN} \tag{11.8}$$

In the price–cost equations (11.1) and (11.2) units for each product type are chosen so that it takes one unit of labour to produce one unit of product. Since by assumption $w_N > w_S$, any product which can be produced in the South, because the method of producing it has already diffused, would be produced only in the South. It follows that n_N is the number of different (types of) products which are sufficiently new that their methods of making are not yet known in the South. Parameter α in (11.3) is the innovation rate and β in (11.4) is the diffusion rate. The utility function assumed in (11.6) is a specification of the constant elasticity of substitution function: it generates positive demands for all the goods that have ever been invented. Finally, (11.8) is the equilibrium condition for the balance of trade.

Provided that L_S and L_N are either constant or change at constant exponential rates, the model has a growth equilibrium solution. The most important part of the solution is that for the ratios of product types and wages:

$$\frac{n_N}{n_S} = \frac{\alpha}{\beta} \tag{11.9}$$

$$\frac{w_N}{w_S} = \left(\frac{\alpha}{\beta} \frac{L_S}{L_N} \right)^{1-\theta} \tag{11.10}$$

Since (11.3), (11.4), and (11.9) imply that $\dot{n}_S$ at any time τ equals n at time $\tau - \alpha/\beta$, the ratio α/β can be interpreted as the technology transfer time-lag. (However, the time which it takes

South to master the production of a new product equals $(1/\alpha)$ $\ln(1 + \alpha/\beta)$.) By (11.9), the transfer time-lag is the ultimate parameter which determines the North–South distribution of all the product types. By (11.10), North's wage advantage is found to be higher the greater is that time-lag and the larger is the South's population in terms of that of the North. The reason is the fact that if the North increases its inventiveness and/or the South increases its share of the world population, the South's import would be higher. In order to limit these imports somewhat and induce exports, so that the balance of trade remains in equilibrium, the South's wages must decline or, alternatively, its imitation power must increase. It is also interesting to see the way in which the elasticity of substitution parameter θ comes into play, together with the ratios α/β and L_S/L_N co-determining the production levels of North and South and the income distribution and trade flows between them. When the goods are perfectly substitutable ($\theta = 1$), the gains from inventing for North and from trading for either party would disappear, as would the North–South real wage differential.

All these implications of the model are intuitively plausible but far from obvious. The Krugman model can also be easily extended to be more realistic, although at a price in terms of simplicity and elegance. One such possible extension would allow for an increasing invention of new ways of making old goods, with the inventions again being diffused worldwide with a time-lag. This process innovation would be another way for the North to defend its wage advantage. This way could also be much more effective, as rapid process innovation in the North may reduce or even eliminate the North–South differences in wage costs per unit of output while sustaining large differentials in wage rates. The presence of such large differentials has in turn two immediate implications: (i) the incentive to transfer production facilities from North to South is usually strong only in cases where production is highly labour intensive but the requirement of highly skilled labour is minor; (ii) income effects are present in the distribution of demand for new products between North and South and consequently in the distribution of production facilities. Since most new products are invented in the North, they are typically designed to meet the more sophisticated demands of a high-income consumer, rather than the basic needs of the South's poorer population. This tends to limit the incentive to set up production in the South to fairly sophisticated goods which can easily be transported to the North's markets or selected simple goods such as minerals.

The disturbingly unrealistic feature of the Krugman model is that the South–North wage differentials would disappear completely if the two regions chose not to trade at all. In the real world it is the intra-North trade which dominates. The South–North trade is too small an affair to be anything but the cause of the wage gap. The North–South trade bias in the Krugman model is thus evident. This bias would matter less if the model were applied to two trading countries which are at a similar level of development, one with a large R&D sector such as the FRG or the USA, and another with a small R&D sector such as Italy. The country in the latter category would be saving R&D resources, but would need to have a somewhat lower wage rate in order to be able to export well-established goods for the purpose of importing recently invented goods.

Licensing, joint ventures, and foreign direct investment

General aspects of technology transfer

In any technology transfer project the main partners are the foreign technology suppliers, the local firm, and the local government. Their separate interests depend in part on the type of the local firm: whether it is national, and then whether it is private or state-owned or foreign, and then whether or not it is a subsidiary (and/or associate) of the technology supplier. From the government perspective, technology transfer is to serve the usual national policy objectives – increasing exports, employment, and the productivity of resources, and doing it at the least cost in terms of imports or inputs, technology transfer payments, and the dependence on foreign economic interests. To promote these objectives, local governments would use a range of measures. In the first instance, laws would be enacted and institutions created to deal with these matters. Then an effort would be made to develop the human resources and social capital (infrastructure) which the use of foreign technology may require. Finally, a fiscal policy would be developed to induce the import of desirable types of technology, which are usually labour intensive, using local natural resources rather than imported inputs and producing import substitutes and/ or exportables. The local government authorities may also try to increase the bargaining power of national firms in their negotiation with foreign suppliers with a view to reducing the technology transfer costs. Developing countries have a poorly developed capital goods sector. The import of these goods is very expensive, and in any case is limited by the ability to export which is typically low. Therefore the authorities in these countries may also place a premium on developing the R&D and production capabilities in the capital goods sector. To this end, in addition to direct grants to that sector, a buy-local policy may be used, implemented by subsidized financing for the purchase of domestic capital goods and by import protection measures.

As we already noted, most developing countries import a very substantial part of all the capital goods that they invest. Both joint ventures and foreign direct investment would typically be based on imported capital goods. Disembodied technology transfers is a separate channel in the sense that it may or may not accompany such imports. The subcategory of commercial transfers of this kind involves specific agreements. These agreements concern the use of patents, trademarks, and technology not protected by industrial property legislation, as well as the supply of technical and engineering services. The preferred form of commercial technology transfer is to some extent influenced by the type of firm using the technology. For example, firms unrelated to the technology supplier must be more dependent on patent and trademark agreements than are its subsidiaries and/or associates. When control over technology or over profits from using it is retained anyway, as in the case of fully owned subsidiaries, the need for formal transfer agreements disappears. This suggests the circumstances when the volume of such agreements underestimates the technology transfer.

Direct foreign investment versus licensing decision

A source country firm (SCF) which is in possession of a technological advantage compared with firms in another country has the option of either licensing that advantage to these firms, itself setting up a host country firm (HCF) through direct foreign investment (DFI), or exporting a product. The problem has been considered by a number of authors, beginning with Hymer (1960, 1976) and Kindleberger (1969) among others. But the models proposed have been rather simple and the topic is still relatively under-researched (Caves 1982: 204–7).

In the initial phase of Vernon's product cycle, the size of the local market is small and therefore exporting the product is then the only choice. When the demand is sufficiently large, one or more local firms prepared to manufacture the product on licence may appear. It is usually only at that stage that the problem arises of choosing between selling a licence or entering the market directly on its own or through a joint venture. Clearly, DFI requires a major commitment of resources now against uncertain profits in the future. For expected profits to be sufficiently high, the size of the market may have to be large for a long period, the degree of competition should be preferably low, and government policies, especially on taxes and profit repatriation, should be attractive and stable. Given the limited local savings, the local authorities may prefer DFI to licensing, on the grounds that the SCF would need to invest resources immediately and be expected continually to upgrade the technology of its HCF. It is clear that the choice between DFI and licensing is situation-specific, depending in particular on the number of possible sources of technology, the structure of the local market, the probable demand for the new product, and the conditions offered by the host government.

Let us consider a very simple case whereby there is as yet no local producer of a product but there is a local firm seeking to establish production on licence. Suppose that the unit variable costs are c^H for the HCF and c^S for the SCF. The respective unit fixed costs would be k^H and k^S. The maximum licence fee that the SCF can obtain is that at which $p = c^H + k^H + f$, where f is the fee per unit of output. At this level of fee, the SCF would capture all the monopoly profit of the local producer. By setting up a subsidiary producing the same output $Q = Q(p)$, the SCF would gain, per unit of output, the difference $(1 - \tau)\{p - (c^S + k^S)\}$ where τ is the rate of tax. The latter option is thus certainly more attractive if

$$(1 - \tau)\{p - (c^S + k^S)\} > p - (c^H + k^H) \tag{11.11}$$

Variable costs of the local product may, but need not, be lower than that of the foreign subsidiary. The latter firm may, for example, employ better quality management and, on foreign markets, make use of the sales network of the parent company. It may also be able to borrow capital at a lower rate of interest than would the local firm. The attractiveness of the direct investment option is increased when the licence fee is to be paid for a period shorter than the expected service life of the plant. However, if there is uncertainty about future prices, tax rates, and exchange rates, or the ability to repatriate profits, then a risk premium must be added to unit costs of the SCF. When this premium is high, the decision may turn in favour of licensing.

Joint ventures – a model due to Svejnar and Smith

A case for setting up a joint venture (JV) arises if all potential partners to it are at least as well off as they would be independently. Such a firm, for an SCF, is a half-way house between just licensing and a full-scale commitment in the form of a fully owned subsidiary. It offers the prospect of profits larger than a licensing fee while at the same time reducing the resource exposure and the risk of hostile reaction on the part of the local government and/or local business community. JVs are particularly popular in countries seeking foreign capital and technology but in which foreign ownership as such is a politically sensitive issue. The governments of these countries would typically impose an upper limit on the SCF's share of the JV's total equity capital.

An instructive model of JVs has been proposed by Svejnar and Smith (1984). A particular feature of this model is the distinction of two (groups of) inputs, call them X and Y, supplied to the JV exclusively by the partners, foreign and domestic respectively, at negotiable (transfer) prices p_X and p_Y. There may also be other inputs, to be denoted by L, but these can be purchased at fixed market prices $\bar{p}_L$. The resource costs of the inputs X and Y, c_X and c_Y, respectively, are assumed constant. The partners may therefore seek to capture profits indirectly through the choice of transfer prices. The accounting profits arising from the JV producing output Q at price p are as follows:

$$\pi = pQ - p_X X - p_Y Y - \bar{p}_L L \tag{11.12}$$

However, the full profits are as follows: for the domestic partner,

$$\pi_D = (p_Y - c_Y)Y + (1 - \sigma)\pi \tag{11.13}$$

for the foreign partner

$$\pi_F = (p_X - c_X - t_X)X + \sigma(1 - t_\pi)\pi \tag{11.14}$$

and for the government

$$\pi_G = t_X X + t_\pi \sigma \pi \tag{11.15}$$

where σ is the (institutionally determined) equity share of the foreign partner, t_x is the import levy per unit of X, and t_π is the profit tax rate for the foreign partner.

The three parties are assumed to distribute total profits in proportion to their bargaining powers. This effectively amounts to assuming that they act as if maximizing, with respect to X, Y, L, p_X and P_Y, a utility function

$$U = \pi_D{}^{\gamma_D} \pi_F{}^{\gamma_F} \pi_G{}^{\gamma_G} \tag{11.16}$$

where γ_D, γ_F, and γ_G are the respective bargaining powers, and $\gamma_D + \gamma_F + \gamma_G = 1$. The corresponding first-order conditions for a maximum give the following results:

$$p\frac{\partial Q}{\partial X} = c_X \qquad p = \frac{\partial Q}{\partial Y} = c_Y \qquad p\frac{\partial Q}{\partial L} = c_L \qquad (11.17)$$

$$\pi_F = \gamma_F \pi^* \qquad \pi_D = \gamma_D \pi^* \qquad \pi_G = \gamma_G \pi^* \qquad (11.18)$$

where $\pi^* = pQ - c_X X - c_Y Y - c_L L$, i.e. the total profit of the JV if its inputs are evaluated at their resource cost. The three conditions in (11.17) are the criteria for the choice of our three (groups of) inputs. It is interesting that the JV would equate the marginal value products of inputs to their unit resource costs rather than to their accounting prices. These prices therefore have no impact on the purchase of inputs. Moreover, the three conditions in (11.18) imply that the profits earned by the three partners are also independent of these prices (as well as of equity shares and the government's tax rates). These results would suggest that the transfer prices and tax rates may not be particularly effective instruments for any of the three parties for the JV to influence their gain from it. What apparently matters most is the bargaining power. Unfortunately, the latter is a rather elusive concept. In practice the bargaining powers of the two business partners, the foreign and domestic firms, are probably related to their equity shares σ and $1 - \sigma$. In any case the model confirms that the comparative advantage of a JV, for the foreign partner, is in acquiring access to inexpensive local inputs and, for the domestic partner, is in acquiring the ability to supply the product at all, and doing so at a profit.

Edwin Mansfield and his research associates established a number of useful facts about the factors influencing, among others, the rate, resource costs, and age of the technology being transferred (Mansfield *et al.* 1982). In particular they report evidence suggesting that US companies tend to transfer newest technology first to their overseas subsidiaries in the developing countries, and only later through licensing or joint ventures. This pattern is consistent with the interpretation that SCFs wish to retain full control over new technology during the first 5–10 years. The resource cost of absorbing new technology tends to be lowest in their subsidiaries, especially those located in developed countries; this factor may therefore also contribute to the observed pattern.

Chapter twelve

Macrotheories and evidence of international technology transfer

Transfer costs and 'appropriate technology'

In any modelling of economic growth that is driven by the import of foreign technology, it is essential to have a good understanding of the variables affecting technology transfer costs. These costs are the resources that are needed in order both to obtain a foreign technology and to ensure its effective utilization. They comprise royalty fees, paid directly or through transfer prices, and the cost of transmitting and absorbing all the relevant know-how, which includes peripheral information relating to plant design and construction, equipment installations, organization and operation of production, quality control, modifying the technology, and training of labour.

A fairly detailed discussion of technology transfer costs is given by Teece (1977) in a case study to which we referred on pp. 197–8. In that particular study, let us recall again, Teece identified four variables as particularly important in influencing the size of the transfer costs: (i) the extent to which the technology is understood by the transferee, (ii) the age of the technology, (iii) the number of firms utilizing the same or a similar technology, and (iv) the cultural and systemic characteristics of the importing country. His econometric estimates imply that 'transfer costs decline as the number of firms with identical or "similar and competitive" technology increases, and as the experience of the transferee increases' (Teece 1977: 253).

Suppose that output Y, capital K, and labour L are related through a constant-returns-to-scale production function $Y = F(K, TL)$, where T is the aggregate index of technology. Let T_h represent the technology level at home, T_F the level in the technology frontier area (TFA) and T_m the level of imported technology (Figure 12.1(a)). Teece's econometric estimates suggest that the international transfer costs rise with T_m, given T_h and T_F, and fall with T_F, given T_h and T_m. Domestic technology transfer costs were also considerable, although in most cases significantly lower than the international transfer costs. The cross-section variation of transfer costs, both domestic and international, was very wide. However, Teece found that 'the most difficult and hence costly technology to transfer is characterised by very few previous applications, a short elapsed time since development and limited diffusion' (Teece 1977: 249). This would imply that the costs of transferring technology T_m are particularly high when $T_F - T_m$ is low, as T_m is then near to, in Teece's terminology, the 'leading-edge' technology, which is as yet relatively untried and in a state of flux. Also royalty fees for such technology can be expected to be higher, as it would be in limited supply and can be thought to offer potentially larger benefits.

We shall need some new notation. Let

$$t_h = \frac{T_h}{T_F} \qquad t_m = \frac{T_m}{T_F} \qquad x = \frac{T_m - T_h}{T_F - T_h} \qquad (12.1)$$

Both t_h and t_m are indices of the relative levels of domestic and imported technology respectively, while x is an index of the novelty of T_m for the importing country. Note that $0 \leqslant t_h, t_m, x \leqslant 1$. Let W represent the total (international and domestic) costs of transferring technology T_m in association with investment projects, the value of which is I, incorporating that technology. Denote

$$R/I = r \qquad (12.2)$$

Given T_h and T_F, the ratio r can be expected to be related positively to x. Later in this chapter we shall consider the implications of a specification which assumes that r is an increasing and strictly convex function of x alone, irrespective of T_h and T_m, as in Figure 12.1(b). This specification has the desired property that r increases with T_m, given T_h and T_F, and falls with T_F, given T_m and T_h. It also implies, and this in some cases may be highly unrealistic, that what matters for transfer costs is the relative rather than the absolute technological gap. Moreover, the unit costs $r(x)$ may in practice be dependent to some extent on the size of the investment flow.

In the discussion above we rely on the conventional distinction between 'technology' and 'technique' which was described in Chapters 1 and 8. Thus technology T_h refers to those among

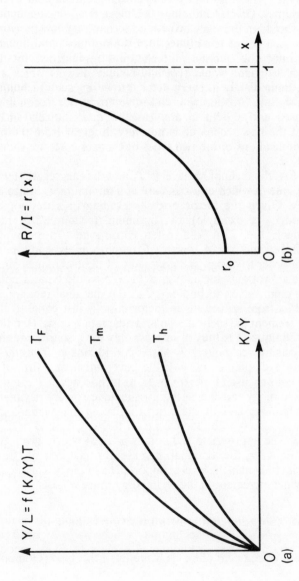

Figure 12.1 (a) Production functions in the backward and modern sectors of a developing country and in the TFA; (b) technology transfer costs per unit of investment in fixed capital as a function of the level of sophistication of the imported technology.

all the techniques known at home which are Pareto efficient. These techniques need not, of course, form a smooth production function of the type $Y = F(K, T_h L)$, with constant returns to scale. However, when they do, a single scalar is enough to represent the technology, while the ratio K/L or K/Y, in addition to the technology index, can be used to identify one of the many techniques. Given technology, a change from one technique to another should in principle involve no technology transfer costs. In practice, changes of techniques of this kind may require some resources, since the alternative techniques represented by the conventional isoquant would typically include designs which are not known immediately in every detail. However, such technique transfer costs are qualitatively different from our technology transfer costs, and should in any case be quantitatively small, especially if the new technique is not very different from the old one. The implications of the two types of transfer costs are shown in Figure 12.2.

In the TFA, the technology level is T_F and the (aggregate) technique used, one at which unit costs are at a minimum, is indicated by point A. Given the factor prices prevailing in a developing country (India, for example), the optimum 'technique' for that country which the TFA can offer corresponds to point C. However, if moving from A towards C requires an expenditure of resources in order to design and adapt any of the techniques other than A, the appropriate techniques to choose would be at a point D. At that point, given technology T_F, 'the marginal increase in social benefits from reductions in factor costs is just equal to the marginal increment in (such) design and adaptation costs' (Findley 1978b). Technique D is thus of interest only for a source country firm which has the technology T_F and is considering an investment project in a developing country. If the government or a firm of a developing country itself is the investor, its technology is T_h. In this case the technology transfer costs must include royalty payments for new technology and/or the cost of imitating that technology. These costs are likely to be increasing on the way from B to C in Figure 12.2. The appropriate technology is T_i, at which, given the ratio K/L or K/Y, the marginal increase in social benefits from any further increment in the level of imported technology is just equal to the marginal increment in the technology transfer costs.

A model of innovation and growth involving technology transfer

An important component of an early model of this type (Gomulka

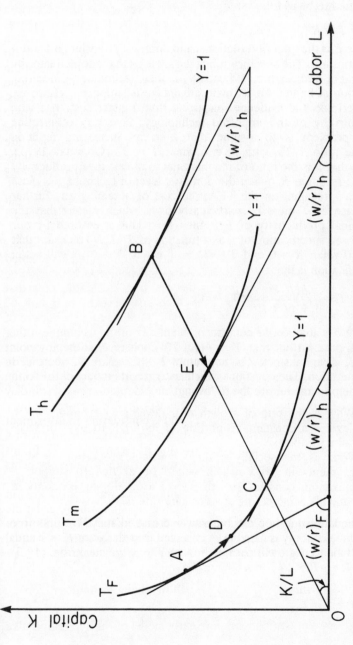

Figure 12.2 The difference between movement from B to E, to reach an 'appropriate' technology, and movement from A to D, to reach an 'appropriate' technique. In both cases point C cannot be reached because of excessive transfer costs.

1970) is the technology transfer cost function. The function is specified to be as follows:

$$\dot{T} = R^{\mu} \tag{12.3}$$

where R is the total (international and domestic) transfer cost and μ is a constant. The specification is the same as that adopted later by Dasgupta and Stiglitz (1980a, b) for their technology-generating function (Chapter 4). Its main advantage is simplicity. However, it overlooks the evidence suggesting that transfer costs rise with the novelty of the imported technology. One way of meeting that problem is to assume that μ is an increasing function of the ratio T/T_F, with $\mu = 0$ when $T = T_F$ (Gomulka 1971). Note that μ is the ratio of the marginal to the average productivity of the resource R. When the domestic level of technology is close to that prevailing in the TFA, the cost of obtaining any further increase in T must be particularly high, which means that the marginal productivity of R is then low. This is ensured by our proposed specification of μ as a function of T/T_F. (To ensure that $\dot{T} = 0$ when $R = 0$ and $\dot{T} = T_F - T$ when $W = \infty$, a still better specification is that

$$\dot{T} = \frac{b(R/I)^{\mu}}{1 + b(R/I)^{\mu}} (T_F - T)$$

where b is any postive constant.)

Suppose further that $Y = F(K, TN)$, where N is employment and K is capital stock. The net output $Y = C + \dot{K} + W$, where C is consumption. The questions which an aggregate model of this kind can help to answer are the following.

(i) What is the rate of growth of a developing economy which devotes a constant proportion of its GDP to pay for technology imports?

(ii) What is the socially optimal constant proportion of such payments in GDP which would be needed to support a maximum constant growth rate of a developing economy in which N is increasing at a constant rate n?

The latter question can be answered immediately. We first note that the economy is capable of balanced growth. Since $\mathring{R} = \mathring{Y}$ and, on a balanced-growth path, $\mathring{Y} = \mathring{K} = \mathring{T} + n$, we have from (12.3) that $\mathring{T} = \mu\mathring{R} = \mu(\mathring{T} + n)$. Hence

$$\mathring{T} = \frac{n\mu}{1 - \mu} \equiv \alpha \tag{12.4}$$

The latter is also the growth rate of labour productivity Y/L and consumption per person C/N. While the capital productivity Y/K remains constant, the ratios I/Y and R/Y can be chosen so that the consumption level is maximized. Since $C = F(K, TN) - gK - R$, the first-order maximization conditions imply that

$$\frac{I}{Y} = a \tag{12.5}$$

and

$$\frac{R}{Y} = (1 - a)\mu \tag{12.6}$$

where $a = (K/Y)\partial F/\partial K$. Hence the ratio of the technology transfer costs to investment is given by

$$\frac{R}{I} = \frac{1 - a}{a}\mu \tag{12.7}$$

From (12.4) and (12.6) it is evident that μ is the key parameter. An increase in μ induces a proportional increase in the ratio R/Y and more then a proportional increase in the innovation rate. If μ is not a constant but a function of the relative technological gap T/T_F, the balanced-growth ratio W/I and the rate α would also be changing with the size of the gap. As argued above, the elasticity μ may be high when $T \ll T_F$, but it should be declining as the technology-importing country comes close to the world's technology frontier. The effect of this decline, according to this theory, would be a fall in both R/Y and α. Once α falls to α^*, which is the innovation rate observed in the TFA, the ratio T/T_F stops increasing further; the developing country would have then reached the best relative position it can obtain through this type of technology transfer. The magnitude of such an equilibrium gap would depend on the precise form of the relationship between μ and the ratio T/T_F. Judging from evidence in Chapter 9, this relationship is country-specific, depending above all on the economic system and cultural factors which jointly influence the speed and effectiveness of the technology transfer process. In an early phase of a developing country's effort to catch up, the technology transferred would be fairly old and therefore inexpensive. Yet private technology transfer costs may be high on account of poor social overhead capital and a limited pool of trained labour. These costs may therefore inhibit the transfer during that phase, keeping the innovation rate low. However, as public investment infrastructure and basic education increase and accumulate, the positive externalities reduce private transfer costs, gradually creating conditions for an innovation take-off.

A major weakness of the model above is its unclear institutional assumptions. Since consumption is maximized, conditions (12.5) and (12.6) can be interpreted as giving 'socially optimal' investments in fixed capital and technology transfer. These conditions may be of interest to central government, but not to individual enterprises. Another problem with the model is the assumption that the capital stock is technologically homogeneous or, at least, that its technological composition remains unchanged. This assumption may be acceptable under the conditions of balanced growth but is clearly unrealistic during the period of catching-up.

The Findley model

The Findley model (Findley 1978a) distinguishes explicitly between domestic and foreign capital, and it was developed to capture the innovation effects of the technology transfer by foreign companies. Using the notation $x = T/T_F$, $y = K_f/K_d$, where K_f is the capital stock imported by foreign companies and K_d is domestically produced capital, we obtain the key equations of the model as follows:

$$\alpha = f(x, y) \qquad f_x > 0 \qquad f_y > 0 \tag{12.8}$$

$$\dot{K}_f = s_f(1 - \tau)\rho_f K_f \tag{12.9}$$

$$\dot{K}_d = s_d(\rho_d + \tau\rho_f y)K_d \tag{12.10}$$

Implicit in (12.8) is the assumption that all technology comes from abroad. The rate of international technology diffusion is higher the greater the relative technology gap and the larger the presence of foreign capital. The model can thus be seen to marry the Veblen–Gerschenkron 'relative backwardness' hypothesis with Mansfield's 'contagion' effect (Chapter 6). The assumption suffers from not recognizing explicitly the existence of transfer costs and from ignoring the fact that these transfer costs may be prohibitively large both in highly backward countries and when the technology transferred is of the frontier type or near to it. In (12.9) ρ_f is the profit rate, τ is the profit tax rate, and s_f is the (constant) fraction of after-tax profits that is retained by foreign companies for net investment in the host country. In (12.10), ρ_d is the profit rate earned by domestic capital and s_d is the fraction of profits and tax revenues that are used to finance its expansion. The profit rate ρ_d is assumed constant, but ρ_f is expected to decline as the ratio T/T_F

increases. The incentive for foreigners to invest is thus greatest when the country is most backward. The assumption is consistent with the hypothesis that there are advantages of backwardness, but it is again unrealistic when technology transfer costs are exceptionally high because the 'absorptive capacity' is low, which is typically the case when the host country is particularly backward.

Since no theory is produced which would enable us to derive α explicitly as a function of x and y, the model cannot be used to trace out the growth path during the catching-up period. However, it can be used to find the size of the equilibrium technology gap and how that size may be responding to changes in the parameters of the model, such as the tax rate or the saving rate. To see this, we note that, at equilibrium, $\alpha = \alpha_F$ and $\dot{K}_f = \dot{K}_d$. Hence the equilibrium x^* and y^* solve the equations

$$f(x^*, y^*) = \alpha_F \qquad (12.11)$$

and

$$s_f(1 - \tau)\rho_f(x^*) = s_d\{\rho_d + \tau\rho_f(x^*)y^*\} \qquad (12.12)$$

These equations give x^* and y^* implicitly as functions of α_F, τ, s_f, s_d and ρ_d. The key results of the model are only qualitative, namely that an increase in the tax rate τ or the savings rate s would increase x^* and decrease y^*, while an increase in α_F or r would have the opposite effect. These results are not particularly surprising. In a model such as this, any change of circumstances which reduces the presence of foreign capital would increase the country's relative technology gap. Thus changes in x^* and y^* go in opposite directions. Since an increase in the tax rate on foreigners' profit reduces the incentive to invest, it would reduce y^* and hence increase x^*. An increase in the saving propensity by domestic investors would increase K_d, hence reducing the ratio K_f/K_d and increasing x^*.

A feature of Findley's model is that foreign companies alone are the venue for channelling foreign capital. In practice, some of the savings generated by the country's own companies are also used to acquire foreign capital goods. In fact in most countries it is this domestic channel that has been and is the most important. Another unrealistic feature is the assumption that only the profits earned by the existing foreign capital can be used to finance its further expansion. However, foreign companies are typically willing and able to borrow in international money markets. Consequently the diffusion process may in practice be faster than implied by the model. A theoretical implication is that in (12.9), r

ɔuld be regarded as a decision variable to be determined rather
an a parameter that is given.

Technological duality and the catching-up

A simple, elegant, and plausible explanation of the catching-up
sequence is provided by making use of the evidence of strong and
persistent duality in technological levels of economic activities in
developing countries.

In any economy, backward or advanced, there is some variation
in the productivity index for inputs among production units.
However, the variation appears to be much larger in the less
developed countries, where two technologically quite distinct
subeconomies can typically be distinguished. A large but stagnant
traditional sector coexists side by side with a small but expanding
(fairly) modern sector. The former comprises mostly agricultural
and services, handicrafts, and small-scale manufacture, while the
latter includes modern farming, most mining, some services, and
most larger-scale manufacturing. To put it differently, the traditional sector includes predominantly activities based on home-made
technology, while the activities of the modern sector are based
largely on the imported technology. A high productivity of the
modern sector finances higher wages and earns high profits, which
in turn attract to that sector more resources of skilled labour and
technology-intensive capital. The economy-wide consequences of
such uneven rates of development are easy to see with the aid of a
simple model.

Let T_h and T_m stand for the average index of technology in the
traditional sector and the modern sector respectively. Also, let α_h
and α_m denote the innovation rates in the two sectors, e.g. $\alpha_h = \hat{T}_h$
and $\alpha_m = \hat{T}_m$. The index of technology is measured by the output
per man-hour, so that the outputs of the two sectors are

$$Y_h = T_h L_h = \frac{K_h}{v_h} \qquad Y_m = T_m L_m = \frac{K_m}{v_m} \qquad (12.13)$$

where L is employment, K is capital stock, and v is the capital-to-
output coefficient. Initially, the economy is dominated by the
traditional sector, so that the average index of technology, equal to

$$T = \frac{T_h L_h + T_m L_m}{L_h + L_m} \qquad (12.14)$$

is close to T_h, and the average rate of innovation α is close to α_h.
Suppose that $\alpha_m = \alpha_F$ and $\alpha_h < \alpha_F$. The country's modern sector

lags behind the TFA, meaning that $T_m < T_F$. The assumption that $\alpha_m = \alpha_F$ implies a constant time-lag, which is plausible as long as the bulk of the technology in that sector is imported.

Since, initially, α is close to α_h, and by assumption $\alpha_h < \alpha_F$, the technological gap between our country and the TFA is increasing. However, if the savings ratio in the modern sector is sufficiently high, employment in that sector would be increasing faster than in the traditional sector. As the modern sector captures more and more resources, the average level of technology will be increasing from T_h to T_m. Thus the relative gap $(T_F - T)/T$ will ultimately be reduced. The speed of that reduction is higher the larger is the investment activity in the modern sector. The innovation rate α would at some point reach the level α_F. At that point the catching-up begins. The maturity stage is reached when the traditional sector disappears, so that $T = T_m$. At that point the catching-up ends, and the (minimum) equilibrium gap is reached.

There are four phases in the development process described above: (a) the initial phase when $\alpha < \alpha_F$, so that the relative backwardness is increasing; (b) a catching-up phase when $\alpha > \alpha_F$ and α is increasing; (c) another catching-up phase when $\alpha > \alpha_F$ but α is declining, which is the Veblen–Gerschenkron case; (d) $\alpha = \alpha_F$, the equilibrium phase.

In our two-sector economy the catching-up is driven by the expansion of the modern sector. The expansion is contingent on the constant supply of foreign technology. Since there are costs of acquiring that technology, its novelty is a choice variable. Earlier in this chapter we suggested relating the unit technology transfer costs to an index of that novelty for the importing country (equation (12.2) and Figure 12.1 (b)).

The full model of the economy can now be written as follows:

$$v_h T_h L_h = K_h \tag{12.15}$$

$$v_m T_m L_m = K_m \tag{12.16}$$

$$v_h \frac{d}{dt}(T_h L_h) = \mathring{K}_h = s_h Y_h - \delta_h K_h \tag{12.17}$$

$$v_m \frac{d}{dt}(T_m L_m) = \mathring{K}_m = s_m Y_m - \delta_m K_m - R_m - R_h \tag{12.18}$$

$$\mathring{T}_m = \mathring{T}_F = \alpha_F \tag{12.19}$$

$$\mathring{T}_h L_h = \alpha_0 T_h L_h + \frac{s}{v_h}(T_m - T_h)L_m \tag{12.20}$$

217

$$R_h = r_h(x)sT_mL_m \tag{12.21}$$

$$R_m = r_m(x)(K_m + \delta_mK_m) \tag{12.22}$$

$$x = \frac{T_F - T_m}{T_F - T_h} \tag{12.23}$$

Since the model is that of a labour-surplus economy, we can assume that there is always enough labour to operate the capital stock. The employment level in each sector is therefore proportional to the sectoral capital stock. This idea is embodied in equations (12.15) and (12.16), themselves implied by the production functions (12.13). Since the economic development is driven by the accumulation of capital and technological change, the key role in the model is played by propensities to invest and by the division of resources between investing in new capital stock and acquiring new technology. In the h sector, for example, the gross investment flow s_hY_h equals the sum of replacement investment δ_hK_h, technology transfer costs R_h, and net fixed capital investment $\dot{K}$. The parameters s_h and δ_h in (12.17) are given, but the unit transfer cost r_h in (12.21) is related to the novelty of the transferred technology x, which is a choice variable.

The model is thus seen to combine international transfer of 'appropriate' technology T_m to the modern sector with an internal diffusion of that technology from the modern to the traditional sector. What is 'appropriate' is endogenously determined by the economic agents who take investment decisions. The decision criterion for the choice of T_m might be profit maximization for private investors operating in the modern sector or accumulated (discounted) consumption for government planners. The effect of internal diffusion is captured by the last term in (12.20), which is a balance equation for the changes in the amount of labour in efficiency units in sector h that are brought about by technological innovation in that sector. The term $\alpha_0 T_hL_h$ is the home-originated change of this kind, and sT_mL_m is the investment flow from sector m to sector h. That investment gives rise to output flow $(s/v_m)T_mL_m$. The latter is equivalent to expressing labour $(s/v_h)L_m$ in terms of efficiency units. If that labour were using technology T_h, its amount in efficiency units would be $(s/v_h)T_hL_m$. The term $(s/v_h)(T_m - T_h)$ in (12.20) is thus the gain in labour units which is brought about by the diffusion to sector h from sector m of technology T_m. On dividing that equation through by T_h, we find that the internal diffusion increases the own rate of innovation of the traditional sector by $(s/v_h)(T_m - T_h)/T_h$. Since the actual note in the modern sector is α_F, but would have been α_0 in the absence of diffusion, the difference $\alpha_F - \alpha_0$ is the international diffusion effect in that sector. The total economy-wide effect is a weighted sum of the two diffusion-related rates, augmented by the composition effect of the labour force moving from the low productivity h sector to the high productivity m sector.

For the international technology transfer to take place at all, the additional capital costs and the transfer costs must not exceed the savings on wage costs. To illustrate the concept of the 'appropriate technology' at a firm level, let us assume that the wage rate w_m in the modern sector is $(1 + a)w_h$, where a represents a premium for higher skills required in that sector in comparison with the traditional sector. Output continues to be related to inputs as in (12.13). Expressing both benefits and costs as a proportion of output, we have that the savings on wage costs are

$$\frac{w_h}{T_h} - \frac{w_m}{T_m} = \left(\frac{1}{T_h} - \frac{1+a}{T_m}\right)w_h$$

the transfer costs are $r(x)v_m$, and the additional capital costs are $i(v_m - v_h)$ where i is the interest rate. Given the level of output, a profit-minimizing firm would wish to choose T_m so as to minimize the net benefit:

$$\varphi = \left\{\frac{1}{T_h} - \frac{1 + a(x)}{T_m}\right\}w_h - r(x)v_m - i(v_m - v_h) \quad (12.24)$$

where a is an increasing function of x, and x is related to T_m by (12.23). For the transfer to take place, there would have to exist a T_m in the range between T_h and T_F such that $\varphi > 0$.

The growth path described by the system (12.15)–(12.23) can be computed numerically. Such computer simulations would help to identify the significance of the different parameters of the model and the forms of the transfer cost functions. As in Findley's model, an analytical solution can be obtained for the situation when the catching-up process has ended, giving way to balanced growth. In such a growth equilibrium both $\hat{T}_F$ and $\hat{T}_m$ are equal to α_F, Consequently, internal diffusion is no longer capable of bringing the technology of the traditional sector closer to that of the modern sector. Moreover, $\hat{L}_h = \hat{L}_m = n$, so that the employment composition remains constant. Other properties of the equilibrium can be derived from the model (12.15)–(12.23). The key equations are as follows:

$$\alpha_F + n + \delta_h = \frac{s_h}{v_h} \quad (12.25)$$

$$v_m(\alpha_F + n + \delta_m) = \frac{s_m - sr_h(x)}{1 + r_m(x)} \tag{12.26}$$

$$\alpha_F = \alpha_0 + \left(\frac{T_m}{T_h} - 1\right) \frac{L_m}{L_h} \frac{s}{v_h} \tag{12.27}$$

Regarding savings ratios as given behavioural parameters and capital-to-output ratios as given technological parameters, there are five unknowns: L_m, L_h, T_m, T_h, and x. However, there are only three equations involving these unknowns: (12.23), (12,26), and (12.27). We close the model by requiring that, in the labour market of each sector, supply equals demand:

$$L_m = \overline{L}_m \qquad L_h = \overline{L}_h \tag{12.28}$$

In a market economy, however, enterprises can be expected to select capital-to-output ratios, as well as s and s_m, with a view to maximizing current profits. The resultant maximization conditions would represent four additional equations to determine v_m, v_h, s_m, and s.

In practice enterprises need not be profit maximizing and there may be some built-in rigidities in the distribution of income. Nevertheless the model above may serve as a useful illustration of the main economic forces at work in the catching-up process. Its particularly distinctive and attractive feature is that the rate of innovation depends on the rate of investment in fixed capital in the modern sector; the greater that investment rate is, the faster is the diffusion of the 'modern technology' in the whole economy. In the 'maturity stage', the backward sector effectively disappears. By that time the country's relative technological gap *vis-à-vis* the TFA declines from the initial $(T_F - T_h)/T_h$ to $(T_F - T_m)/T_m$. To reduce this gap further, the country would need to develop its own inventive activity or become the host of companies of the TFA.

Import-led growth as a policy

Import-led growth is a term reserved for a particular form of technology transfer, one in which imported technology and perhaps other productive assets are used to increase outputs first and to pay for these imports only later, with a fraction of the additional outputs which can be directly or indirectly traced to the imports. The policy assumes that outputs are limited by resource constraints and, more importantly, that the cumulative gain in outputs will be sufficiently high to leave a surplus after the payment for imports, any interest for credit, and any domestic production and transfer costs. At an economy-wide level, import-led growth is thus a policy whereby there is initially a net inflow of resources, new technologies in particular, and in which, if the policy is successful, there is an acceleration of growth in both the short and long term, despite a net outflow of resources in the medium term.

Many newly industrialized countries in Latin America, the Far East and Eastern Europe have adopted the policy in the 1960s and especially in the 1970s. In some of the countries, particularly in the Far East, the policy was apparently successful. However, in much of Latin America and Eastern Europe large credit imports led to the well-known debt management problem. Attempts to deal with the problem through import cuts and export promotion in the 1980s led in turn to large-scale recessions in some of the more indebted countries.

The experience has been the subject of detailed country studies, many of them initiated by the World Bank and the OECD. Two reasons for the failure of the policy stand out as crucial. One is an unexpectedly large rise in interest rates above historical levels in the late 1970s and during the 1980s. The other is the low capacity of many of the importing countries to make good use of the credit-financed imports, in particular their failure to develop the export sector fast enough.

A number of studies attempted to test the usefulness of the policy by seeking to evaluate the productivity of imported capital in comparison with that of domestic capital. Gomulka and Nove (1985) have surveyed these studies with reference to the USSR and Eastern Europe. The estimation results proved uncertain, depending in part on the specification of the production function adopted by the various studies. The specifications of the function usually tried were the following:

$$Y = K_i^a K_h^b L^c \exp(\lambda t) \tag{12.29}$$

or
$$Y = (K_i + \omega K_h)^a L^c \exp(\lambda t) \tag{12.30}$$

or
$$Y = \{\delta K_i^{\xi} + (1 - \delta)K_h^{\xi}\}^{1/\xi} L^c \exp(\lambda t) \tag{12.31}$$

$$Y = F\{K_i, \omega L_i \exp(\lambda t)\} + F\{K_h, L_h \exp(\lambda t)\} \tag{12.32}$$

where K_i is the imported capital and serves as a proxy for technology imports, K_h is home-produced capital, L is employment, and Y is net output, i.e. value added reduced by the cost of investment and technology imports.

221

In (12.29) the imported capital is an essential input, so that $Y = 0$ if $K_i = 0$. This assumption tends to exaggerate the contribution of K_i. Another flaw is that the innovation rate λ is independent of the size of technology imports. In (12.30) K_i and K_h are assumed to be perfect substitutes. The purpose is to find ω which measures the technological advantage of the imported capital. In (12.31) the substitution elasticity between K_h and K_i is a parameter to be estimated. Specification (12.32) corresponds to our duality model given above. Its desirable feature is that the aggregate rate of technological change would be related to the size of technology imports. However, it has a practical disadvantage in that the data L_i or L_h on employment are seldom available.

In our dual-economy model we distinguished between international technology transfer and domestic technology transfer. The latter arises when investment goods produced in the modern sector themselves incorporate imported technology. By producing such investment goods, the modern sector in effect increases the volume of imported capital, contributing indirectly to the spread of its technology. The two forms of transfer can therefore be called, respectively, the direct and indirect forms of international technology diffusion (Gomulka and Sylwestrowicz 1976). The model (12.15)–(12.23) incorporates the innovation and reduces the output effects of the two forms, while the specification (12.32) incorporates only the effects of the direct form. An empirical attempt to estimate the effects of both forms (Gomulka 1976) suggests that the indirect form of diffusion may, under some circumstances, be more important than the direct form. Our duality model can be used to simulate growth paths under plausible assumptions on technology transfer costs. Such simulations could assist in evaluating the contributions of the two forms of transfer during the catching-up phase.

In the analysis of the possible growth effects of international technology transfer we may also wish to recognize the vintage structure of the capital. In particular, if technological change is of the embodied type, new technology imports can affect the productivity of the existing capital stock only. This restriction reduces the output effect of such imports and therefore the incentive to undertake the policy of import-led growth. (For a recent attempt to model international technology transfer in the framework of a vintage-type model see OECD (1989).)

Chapter thirteen

Innovation rate and change of economic systems: a grand scenario

Major historical trends: a summary

Nelson (1981: 1035) notes that much of the research work in the field of growth and technological change, has been concerned to explore and answer the following three broadly separable questions. What lies behind a particular country's growth rate and its variation over time? What explains differences in levels and rates of change of productivity among countries? Why do certain firms and industries experience much faster productivity growth than others? In this book, we have also sought to provide answers to these central questions. It may be useful to summarize these answers.

The key initial methodological step in our whole inquiry was the supplementary question: what would output and productivity growth – of a particular industry, country, or the world as a whole – have been in the total absence of any innovation over a long period of time? The purpose of this question was to separate out the proximate from the ultimate factors of growth, or the short-term from the long-term factors. It was this question which led us immediately, in Chapter 1 to give the pride of place, in long-term productivity growth, to factors that lie behind the production and spread of knowledge and innovation. Standard growth accounting of the Abramovitz–Solow–Denison type has not been rejected in any way, provided that it is applied only for periods of time short enough to ignore the positive feedback effect of innovation, through output and productivity growth, on capital accumulation. For longer periods there is risk of the results of applying the method being misinterpreted and the contribution of capital growth being given too much weight. At the same time, we must and do recognize that the level of investment activity may affect the innovation rate. When that share of investment in output is small, many useful innovations cannot be implemented. The positive relationship between innovation rate and investment share may therefore be expected to be particularly strong for low values of the share. We also came across this relationship in our technological duality model in Chapter 12. However, for this relationship to operate there must be a sufficient supply of both inventions and enterprises eager to make effective use of the inventions. Most investment shares are in the range 15–25 per cent, and so they present a relatively small fraction of total output. Increasing the share from one level to another in this range is therefore feasible for practically all countries, especially if such an increase is implemented gradually and if it leads to a higher innovation rate and sustained productivity growth. It is therefore again the supply of inventions, and the incentives and absorptive capacities which underlie the demand for inventions, that are really crucial and that lie behind a particular country's productivity growth rate.

Given this preliminary view, our major effort has been directed to explaining the observed wide variation in innovation and productivity growth rates over time and among countries. In Part I, which deals with the invention–innovation–diffusion processes at the level of firms and industries, we have sought to develop the microfoundations for the macro-oriented analysis of Part II. Central to that analysis are two hat-shaped or bell-shaped relationships. The Second Hat-Shaped Relationship applies for countries of the technology frontier area (TFA). It is intended to describe the dynamics of the innovation rate, common for all such countries, over time. The First Hat-Shaped Relationship applies to the countries outside the TFA. The empirical basis for this distinction between the TFA and the rest of the world is the extremely high concentration of the R&D and inventive activity in the few most developed countries of the world. Therefore it must be this activity which at present drives the growth processes in the TFA, and it must be the international diffusion of the TFA's technology which largely drives, or is capable of driving, the growth processes in the countries outside the TFA.

The Second Hat-Shaped Relationship, as derived in Chapter 10, is essentially the consequence of the TFA's making an increasingly more effective use of what is perhaps the major natural resource, the research and inventive talents of the world's population. Given the very limited access to education by the children of the poor, this resource, until the nineteenth century, was largely wasted even in Western Europe. Judging by modern experience, suppose that some 3 per cent of the working population can be considered to have sufficient talent to be regarded as potential inventors. This percentage defines the talent pool, the ultimate size of the intellectual resource, or the upper limit for employment in the inventive activity. Since initially only a small fraction of that pool was effectively used, there exists a large scope for a fast growth of that activity over a period of time until the upper limit is reached. This limit is reached first in the TFA. This may already have happened. However, if we take the TFA to mean Western Europe, North America, and Japan, the area accounts only for some 15 per cent of the world's population. There are grounds for believing that a large proportion of the talent pool among the other 85 per cent of the population remains severely under-utilized. A fairly fast growth in world inventive activity should therefore continue for some time yet. For how long? The answer depends, in part, on what we believe that the growth rate of the number of inventors has been in the TFA during, say, the last 100 years. Suppose that this rate has been 5 per cent. The total world population in 1990 is about 5 billion, of which 15 per cent, or 750 million, live in the TFA. Assume further that the population of the present TFA will remain constant, but the population of the rest of the world will be increasing at an annual rate of 1 per cent. Our question is this: assuming that the proportions of potential inventors in the two populations are the same, for how long will the total world employment of inventors be able to continue to increase at 5 per cent if effective unemployment rates among potential inventors are, at present, zero in the TFA and 100 per cent in the rest of the world? The answer is: only about 50 years. The period would be higher if the actual growth rate of employment in the world inventive activity were taken to be less than 5 per cent. However, since the effective unemployment rate among all potential inventors outside the TFA is less than 100 per cent, the period needed to bring about full employment would be shorter. Whatever the length of the period, however, once it has ended, the full utilization of the world total pool of inventive talent is reached, and R&D employment growth must subsequently fall. In the numerical projection above, the fall will be from 5 per cent to less than 1 per cent, the projected growth rate of the world population. Given our understanding of the relationship between this rate and the innovation rate, the latter would begin to fall as well. The further consequences would be a slow-down in productivity growth and in output growth in the TFA.

Several factors may influence this innovation and growth slow-down. One is related to the time-lag between invention and innovation. In Chapter 3 we noted that, in the course of the last two centuries, this time-lag has been apparently declining, at least for the group of major innovations. Such a decline increases the innovation rate, but only temporarily; the gain withers away as soon as the time-lag no longer falls. Similarly, in Chapter 6 we noted that diffusion rates for specific innovations are now generally higher than they were in the past. The resulting gain to the aggregate rate of innovation would again decline and eventually disappear, as it inevitably becomes increasingly difficult to reduce diffusional time-lags further. Moreover, although we did not discuss this topic in the book, it must be noted that productivity gains from technological change would begin to decline if, owing to limited world stocks of physical resources such as land and water, diseconomies of scale started to set in more often. The slow-down in the growth rates of productivity and output may consequently be stronger than the slow-down in innovation rates.

Our second key relationship, the First Hat-Shaped Relationship, is related to the innovation and growth effects of the international technology transfer. It offers the late-comers in industrialization the prospect that their catching-up with the TFA may be in large measure successful. However, despite large and increasing international technology transfer the cross-country variation in technology and productivity levels has not increased much during the last two centuries. This can be seen as a paradox. Its explanation has to do with two facts. One is that for a considerable period of time only a small proportion of the world population, essentially West Europeans, have been exceptionally successful in inventing. The other is that, during that period, international diffusion rates were low. Given the large cross-country variation in institutional arrangements, cultural traditions, and natural environments, it is perhaps not surprising that 'technological revolution' began and continued for some time only in one of the many parts of the world. That part happened to be large enough to sustain it, and the cultural and other barriers were large enough to keep the other parts unreceptive. Only when the technological superiority of the TFA became apparent, and recognized by the 'rest of the world' as a threat or, if taken advantage of, as an opportunity to make economic gains, have diffusion rates increased.

It follows from Figure 9.1 that, in the 1960s, the countries with GDP per capita less than 5 per cent of the US per capita level accounted for about half of the world population. During that decade their relative incomes declined further. However, this was the decade of the Cultural Revolution in China. During the 1950s and since the mid-1970s the economic development in China has been fast. In fact, from 1952 to 1982, per capita national income grew at an average annual rate of 4.0 per cent, accelerating to 6.8

per cent from 1978 to 1984 (*China: A World Bank Country Economic Report* 1985: 1). The level of this income (in terms of purchasing power) stood at approximately a tenth of the TFA's level. The income in India is about the same (or somewhat lower). Suppose that the combined per capita income of China and India increases from 1984 at an annual rate of 4.5 per cent and that, in the TFA, it increases at a rate of 2.0 per cent. Suppose also that the equilibrium gap is reached when the level in China and India equals 50 per cent of the TFA's level. When will this level be reached? The answer is 66 years after 1984, i.e. 2050. Essentially the same forecast is obtained if, owing to the anticipated rapid depletion of non-reproducible resources, the growth rates will be somewhat lower, say 1.0 per cent in the TFA and 3.5 per cent in China and India.

The purpose of this illustrative growth projection is to make the point that the still very pronounced technological and income duality of the world economy today is probably going to be sharply reduced by the middle of the twenty-first century. The First Hat-Shaped Relationship will become largely irrelevant by then.

The international growth equilibrium of the future entails the presence of some variation in productivity levels in the world, of the order observed now in Europe, but there should no longer be much variation in innovation rates and productivity growth rates. Our growth projection exercise also suggests that the world pool of potential inventors may well be near to full employment by 2050. This implication is thus consistent with our earlier illustrative projection, namely that the world pool of fully employed inventors may well grow at the old high rate, taken to be 5 per cent, for another 50–60 years. The two projections are linked: the slower are China, India, and Latin America in their catching-up effort, the lower will be the contribution of their R&D sectors to the world inventive activity and therefore the slower the innovation rate of the TFA.

A global industrialization phenomenon and the transition from pre- to post-industrial society

In our technological change and growth models industrial sectors do not appear to be more important than agriculture or services. The distinctions which we stressed particularly strongly are very broad: between research–inventive–innovative activity and the production of conventional goods, between users of new products and processes and their suppliers, between modern sectors and traditional sectors, etc. Nevertheless, the technological revolution is often associated with definite changes in the sectoral composition of conventional goods, with the agriculture sector giving way to mining, manufacturing, and construction, and these three industries giving way to services. In the industry sector itself there is a shift of employment from heavy manufacturing to electronics and microelectronics (this has recently occurred in the TFA), and in the services sector there is rapid growth of education, health, and information processing. The recent fall in industrial employment in countries of the TFA supports the view that global industrialization is just an instrument of transition for the world population from cultivating land in small and primitive village settings to providing a wide range of services in large and complex urban centres. Since innovations very often have the form of new machinery and buildings or require these goods to be implemented, it is not surprising that the fast growth of inventive activity both caused and was made possible by the rise of industries and a corresponding fall in agriculture. What are perhaps less obvious are the reasons for the recent phenomenon of de-industrialization. When considering the causes of these or any other sectoral change, it is useful to keep in mind that, in the absence of innovation of any kind and assuming constant returns to scale, incomes per head and relative prices would be constant and therefore the product composition of output would be stable. It follows that changes as such in the industrial composition of output and employment must in large part be a consequence of innovation. However, any particular change in that composition, such as from food to consumer durables or from consumer durables to health services, must be dependent on consumer preferences, which although influenced by innovation are best regarded as intrinsic to human nature and non-economic.

Rapid innovation epoch and the capitalization of feudalism and socialism

In Chapters 4 to 6 we discussed, among other things, how the amount of technological change, at the levels of enterprise and industry, can be influenced by a given 'market structure' of industry, and how the market structure itself may be influenced by opportunities to innovate and the incentive to take advantage of these opportunities. However, market structure is just one of many institutional characteristics of an economic system. Technological innovation may lead to such new characteristics, and new ways of organizing and conducting economic activities, i.e. to systemic changes. Equally, any given economic system may influence the innovation rate, which is the subject we discussed in Chapters 7 and 9. In these concluding sections of the book, we return again, if only briefly, to the broad but fascinating theme of the implications for the dynamics of economic systems of the 'technological revolution' and the way we expect this revolution to end.

Three major systemic changes are of particular interest: (i) the transition from feudalism to capitalism which already took place; (ii) the (partial) re-introduction of capitalism, as a result of reforms that are now under way in the (socialist) countries of Eastern Europe, the USSR, and China; (iii) possible gradual socialization of the predominantly capitalist economy of the world in the future.

The transition from feudalism to capitalism is best documented and probably easiest to interpret. It has been driven by industrial innovation on the supply side and a rapid expansion of consumer demand for newly invented non-agricultural products. The drive was sustained by a fast-increasing pool of inventors able to supply an increasing flow of inventions and a sufficient number of entrepreneurs ready to recognize the profitable opportunities that these inventions were creating. Once these were in place, the rest would almost inevitably follow: an income gap would develop in favour of the urban worker, inducing migration from the country-side, and a high profit margin on non-agricultural investments would fuel urban capital accumulation. With time, the number of entrepreneurs and their businesses became large enough to create reasonably competitive markets. These markets, private enterprise, the real prospect of high profits for the inventor and the entre-preneur, and the real threat of poverty for the unsuccessful have emerged as the key features of the capitalist system.

In the early years of capitalism it had already become apparent that the system, if it is to be successful, both causes and requires large inequalities between people in terms of income, wealth, and status, the latter in part related to decision-making powers. At the same time, the system was proving itself to be well designed to induce invention and innovation, and therefore to be capable of rapidly improving the material standard of living of the majority of the people. The following two questions have immediately arisen. Would large inequalities always be acceptable for this majority? Would the system maintain this ability to perform well as it matures?

Both Marx and Schumpeter argued that, in fact, capitalism would give way to socialism. It is interesting that, although their arguments are quite different, they rest on particular assumptions concerning the innovation process. These arguments have so far proved to be in large measure wrong (Marx) or doubtful (Schumpeter). Innovations have indeed been largely labour saving, as Marx assumed, but, contrary to his expectations, they have not caused the increasing and eventually massive unemployment which was supposed to lead to a super-economic crisis and, eventually, the overthrow of the capitalist system. Marx argued that capitalists would be very successful in the pursuit of capital accumulation. This argument proved correct. However, this capital accumulation was supposed to lead to declining rates of profit, which did not occur. We might say that, in Marx's view, capitalism would fall under the weight of its successes in innovation and capital accumulation. In his theory of growth, he clearly underestimated the effect of innovation on wage increases and the effect of increasing wages on demands for goods and the employment of labour. For Schumpeter, the capitalist economic system was one in which the individual initiative of the entrepreneur, rather than the collective efforts of organizations, was central. However, his economies-of-scale argument which we discussed in Chapter 4 led Schumpeter to believe that small firms, in their inventive and innovative activity, would be at a disadvantage compared with large firms, and so the latter would eventually dominate. Firms would increasingly be so large and complex that they would have to be run by hierarchical organizations. A bureaucratized econ-omic system would emerge: 'an order of things which it will be merely a matter of taste and terminology to call Socialism or not' (Schumpeter 1928). Such a system would be less competitive and could in due course become less innovative than the initial capitalist system. Thus for Schumpeter the direction of causality runs from entrepreneurial capitalism to increasing concentration of production and bureaucratization of management (which at some stage he would be prepared to call socialism), and only then to a possible innovation slow-down. So far, however, the nature of the innovation process appears to have been different. The innovation activity and economic significance of small-scale firms continue to be large. Moreover, both advances in information technology (making market information more readily available) and the opening of national borders for trade have tended to sustain a higher degree of competition, especially for internationally traded goods, despite the emergence of super-large firms.

The innovation limits to growth and the socialization of capitalism

The growth of (full-employment) output has always been

constrained by the rate of innovation, in addition to the growth of other resources, producible or non-producible. The concept of conventional limits to growth, however, highlights the fact of limited reserves of non-reproducible resources. The limits begin to bite when the marginal costs of recovering these resources, such as metals or fossil energy, begin to increase rapidly. Unless suitable substitutes are found or invented, decreasing returns to scale set in, reducing the growth rate of output. Inventors are a reproducible resource. However, if, as we suggested in this book, the increasing use of the pool of potential inventors underlies the fast growth during the 'technological revolution', the full utilization of this pool world wide, once reached, imposes a new constraint on the innovation rate. The innovation flow may still be increasing, but at a lower rate. The technology and productivity levels may, as a result, continue to increase, but their growth rates should be declining, possibly to zero (Second Hat-Shaped Relationship, stages (iv) and (v)).

The conclusion above is, for now, only a theoretical possibility. Nevertheless, it may be useful to explore briefly its probable far-reaching 'systemic' implications, if only for the purpose of bringing them to our awareness.

Our theory of innovation and growth suggests an evolution of the world economy, in the course of the twenty-first century, towards a situation in which (i) the flow of innovation is high, but the percentage rates of innovation and productivity growth are low and similar among countries, and (ii) technology and productivity levels are high and continue to vary among countries, but much less than at present. The combined effect of stable incomes per head and slow innovation should be less (percentage or relative) change in the economy. Both demands and relative prices would become less volatile. The job security of workers would increase. The risk for managers to make mistakes and the scope to make large profits would both decline. In a less-changing nearly stationary economy, the entrepreneurial functions of managers would also diminish. The market value of their work, as well as that of inventors, would therefore decline. Market competition would still be needed to induce efficiency. Economic information may date less rapidly, but the information-updating role of markets would continue.

If these were the circumstances of the world economy today, the pressure for centrally managed socialist economies to adopt reforms which make them 'capitalist without capitalists' would probably be much less, for, as we noted in Chapter 7, a major problem of these economies is extreme inertia: an inability both to produce fast change, in particular technological change, and to

231

respond quickly and adequately to change. The link between innovation and income inequality goes in both directions: high rewards are needed as an incentive for rapid innovation, and the gradual introduction on markets of many new consumer goods, initially always in limited quantities, requires differentiated incomes to create correspondingly limited demands. However, to take an extreme case in a completely changeless capitalist economy, there would be, in such an economy, less justification for keeping a small minority of managers, capital-owners, and market-makers in privileged positions in terms of incomes, wealth and decision-making power. Consequently, if innovation slows down, pressure may build up among the working (and voting) population of the world's capitalist countries to adopt 'socialist' tax and ownership reforms which would lead to greater equity. These reforms may even go far enough to amount to some significant 'socialization of capitalism'. Any such socialization may well imply a loss of economic efficiency and therefore lower consumption, but it may well be seen by the majority of the voting population as an improvement in terms of their welfare.

If our prediction of major innovation slow-down proves correct, and if this entails an emergence and world wide spread of a socialist system of some kind, Marx would be vindicated in his belief that socialism is the right economic system only for well-developed countries and only when the development capabilities of capitalism have been exhausted. However, in our theory the extraordinary development capabilities occasioned by the 'technological revolution', and probably best exploited under capitalism, cannot be revived under any post-capitalist system. Socialism, if this proves to be that system, will have been adopted not because of its economic advantages, but because its disadvantages will be less costly.

232

Bibliography

Abramovitz, M. (1979) 'Rapid growth potential and its realisation: the experience of capitalist economies in the postwar period', in E. Malinvaud (ed.) *Economic Growth and Resources*, Vol. 1, *The Major Issues*, London: Macmillan, Chapter 1.

Adams, W.J. (1970) 'Firm size and research activity: France and the United States', *Quarterly Journal of Economics* 84: 386–409.

Ahmad, S. (1966) 'On the theory of induced innovation', *Economic Journal* 76: 344–57.

Aigner, D.J., Lovell, C.A.K., and Schmidt, P. (1977) 'Formulation and estimation of stochastic frontier production function models', *Journal of Econometrics* 6 (1): 21–37.

Alchian, A.A. (1950) 'Uncertainty, evolution and economic theory', *Journal of Political Economy* 58: 211–21.

Aliber, Z.A. (1970) 'A theory of direct foreign investment', in C.P. Kindleberger (ed.) *The International Corporation*, Cambridge, MA: MIT Press.

Ames, E. (1961) 'Research, invention, development and innovation', *American Economic Review* 51 (3): 370–81.

Amman, R. and Cooper, J.M. (eds) (1977) *Industrial Innovation in the Soviet Union*, New Haven, CT: Yale University Press.

Amman, R., Cooper, J.M., and Davies, R.W. (eds) (1977) *The Technological Level of Soviet Industry*, New Haven, CT: Yale University Press.

Arrow, K.J. (1962a) 'Economic welfare and the allocation of resources for inventions', in R.R. Nelson (ed.) *The Rate and Direction of Inventive Activity*, Princeton, NJ: Princeton University Press.

Arrow, K.J. (1962b) 'The economic implications of learning by doing', *Review of Economic Studies* 29: 155–73.

Arrow, K.J., Chenery, H.B., Minhas, B.S., and Solow, R.M. (1961) 'Capital–labour substitution and economic efficiency', *Review of Economics and Statistics* 43 (3).

Atkinson, A.B. and Stiglitz, J.E. (1969) 'A new view of technological change', *Economic Journal* 79: 573–8.

Axelrod, R. (1984) *The Evolution of Cooperation*, New York: Basic Books.

Azam, J.P. (1980) 'The slow economic growth of the UK since the World

Bibliography

War II: a comparison with France', Ph.D. Thesis, London School of Economics.

Baily, M.N. (1981) 'The productivity growth slowdown and capital accumulation', *American Economic Review* 71: 326–31.

Baily, M.N. and Chakrabarti, A.K. (1985) 'Innovation productivity in U.S. industry', *Brookings Papers on Economic Activity* 2: 609–32.

Baldwin, W.L. and Scott, J.T. (1987) *Market Structure and Technological Change*, Harwood Academic Publishers.

Baldwin, W.L. and Scott, J.T. (1989) 'Market structure and technological innovation', in F.M. Scherer (ed.) *The Economics of Technical Change*, Harwood Academic Publishers.

Baranson, J. (1970) 'Technology transfer through the international firm', *American Economic Review* 60 (May).

Barzel, Y. (1968) 'Optimal timing of innovations', *Review of Economics and Statistics* 50: 348–55.

Baumol, W.J. (1986) 'Productivity growth, convergence and welfare: what the long-run data show', *American Economic Review* 76: 1072–85.

Baumol, W.J. and Stewart, M. (1971) 'On the behavioural theory of the firm', in R. Morris and A. Wood (eds) *The Corporate Economy: Growth Competition and Innovative Potential*, London: Macmillan, Chapter 5.

Baumol, W.J. and Wolff, E.N. (1983) 'Feedback from productivity growth to R&D', *Scandinavian Journal of Economics* 85 (2): 147–57.

Baumol, W.J. and Wolff, E.N. (1984) 'On interindustry differences in absolute productivity', *Journal of Political Economy* 92: 1017–34.

Beamish, P.W. (1988) *Multinationals and Joint Ventures in Developing Countries*, London: Routledge.

Beardsley, G. (1977) 'Social and private rates of return from industrial innovations', *Quarterly Journal of Economics*, May.

Beath, J., Katsulacos, Y., and Ulph, D. (1990) 'Strategic innovation', in M.A.H. Dempster (ed.) *Mathematical Models in Economics*, Oxford: Oxford University Press.

Beckerman, W. (ed.) (1979) *Slow Growth in Britain: Causes and Consequences*, Oxford: Oxford University Press.

Benvignati, A.M. (1982a) 'Interfirm adoption of capital goods innovations', *Review of Economics and Statistics* 64: 330–5.

Benvignati, A.M. (1982b) 'The relationship between the origin and diffusion of industrial innovation', *Economica* 49: 313–23.

Bergson, A. (1963) *The Real National Income of Soviet Russia Since 1928*, Cambridge, MA: Harvard University Press.

Bergson, A. (1979) 'Notes on production function in Soviet postwar industrial growth', *Journal of Comparative Economics* 3: 116–26.

Bergson, A. (1983) 'Technological progress', in A. Bergson and H.S. Levine (eds) *The Soviet Economy: Towards the Year 2000*, London: Allen & Unwin.

Berliner, J.S. (1976) *The Innovation Decision in Soviet Industry*, Cambridge, MA: MIT Press.

Bernstein, J.I. and Nadiri, M.I. (1988a) 'Interindustry R&D spillovers, rates of return, and production in high-tech industries', Working Paper 2558, National Bureau of Economic Research.

Bernstein, J.I. and Nadiri, M.I. (1988b) 'Rates of return on physical and R&D capital and structure of production process: cross section and time series evidence', Working Paper 2570, National Bureau of Economic Research.

Binmore, K. and Dasgupta, P. (eds) (1986) *Economic Organisation as Games*, Oxford: Basil Blackwell.

Binswanger, H.P. (1974a) 'A micro economic approach to induced innovation', *Economic Journal* 84 (December).

Binswanger, H.P. (1974b) 'The measurement of technical change biases with many factors of production', *American Economic Review* 64: 964–76.

Binswanger, H. and Ruttan, V. (eds) (1978) *Induced Innovation: Technology, Institutions and Development*, Baltimore; MD: Johns Hopkins University Press.

Bornstein, M. (1986) *East–West Technology Transfer: USSR*, Paris: OECD.

Boulding, K.E. (1981) *Evolutionary Economics*, California.

Boylan, M.G. (1977) 'The sources of technological innovations', in B. Gold (ed.) *Research, Technological Change in Economic Analysis*, New York: Lexington Books, Chapter 5.

Branch, B. (1974) 'Research and development activity and profitability: a distributed lag analysis', *Journal of Political Economy* 82: 999–1011.

Brown, L.A. (1981) *Innovation Diffusion*, New York: Methuen.

Bruno, M. (1978) 'Duality, intermediate inputs and value-added', in M. Fuss and D. McFadden (eds) *Production Economics: A Dual Approach to Theory and Applications*, Amsterdam: North-Holland, Vol. 2, pp. 3–16.

Bruton, H.J. (1967) 'Productivity growth in Latin America', *American Economic Review* 57 (December).

Buckley, P.J. and Casson, M. (1976) *The Future of the Multinational Enterprise*, London: Macmillan.

Buckley, P.J. and Casson, M. (1981) 'The optimal timing of a foreign direct investment', *Economic Journal* 91 (March).

Buckley, P.J. and Casson, M. (1985) *The Economic Theory of the Multinational Enterprise*, London: Macmillan.

Burmeister, E. and Dobell, A. (1970) *Mathematical Theories of Economic Growth*, New York: Macmillan.

Cain, L. and Paterson, D. (1981) 'Factor biases and technical change in manufacturing: the American system, 1850–1919', *Journal of Economic History* 41: 341–60.

Carter, C.F. and Williams, B.R. (1957) *Industry and Technical Progress: Factors Governing the Speed of Application of Science*, London: Oxford University Press.

Casas, F. (1970) 'The theory of intermediate products, technical progress, and the production function', *International Economic Review* 11 (2): 179–208.

Caves, D.W. and Christensen, L.R. (1980) 'The relative efficiency of public and private firms in a competitive environment: the case of Canadian railroads', *Journal of Political Economy* 88 (5): 958–76.

Caves, R.E. (1971) 'International corporations: the industrial economics of foreign investment', *Econometrica* 38: 1–27.

Caves, R.E. (1982) *Multinational Enterprise and Economic Analysis*, Cambridge: Cambridge University Press.

Chang, W. (1972) 'A model of economic growth with induced bias in technical progress', *Review of Economic Studies* 39: 205–21.

Chesnais, F. (1988) 'Technical co-operation agreements between firms', in *STI review*, Paris: OECD.

Chilosi, A. and Gomulka, S. (1974) 'Technological conditions for balanced growth: a criticism and restatement', *Journal of Economic Theory* (October): 171–84.

China: A World Bank Country Economic Report (1985), Washington, DC: World Bank.

Christensen, L.R., Jorgenson, D.W., and Lau, L.J. (1973) 'Transcendental logarithmic production frontiers', *Review of Economics and Statistics* 28–45.

Christensen, L.R., Cummings, D., and Jorgenson, D.W. (1980) 'Economic growth, 1947–73: an international comparison', in J.W. Kendrick and B.N. Vaccara (eds) *New Developments in Productivity Measurement and Analysis*, Chicago, IL: University of Chicago Press.

Christianson, G.B. and Haveman, R.H. (1981) 'Public regulations and the slowdown in productivity growth', *American Economic Review* 71: 320–5.

Chudnovsky, D., Nagao, M., and Jacobson, S. (1984) *Capital Goods Production in the Third World: An Economic Study of Technical Acquisition*, London: Frances Pinter.

Clark, J., Freeman, C., and Soete, L. (1981a) 'Long waves and technological developments in the 20th century', *Konjunktur, Krise, Gesellschaft* 25.

Clark, J., Freeman, C., and Soete, L. (1981b) 'Long waves, inventions, and innovations', *Futures* 13 (4).

Coase, R.H. (1937) 'The nature of the firm', *Economica* 4.

Comanor, W.S. (1965) 'Research and technical change in the pharmaceutical industry', *Review of Economics and Statistics* 47: 182–90.

Comanor, W.S. (1967) 'Market structure, product differentiation, and industrial research', *Quarterly Journal of Economics* 81: 639–57.

Comanor, W.S. and Scherer, F.M. (1969) 'Patent statistics as a measure of technical change', *Journal of Political Economy* 77: 392–8.

Cooper, C. and Clark, J. (1981) *Investment, Technological Change and the Level of Employment*, Brighton: Wheatsheaf.

Cooper, J.M. (1982) 'Innovation for innovation in Soviet industry', in R. Ammann and J.M. Cooper (eds) *Industrial Innovation in the Soviet Union*, New Haven, CT: Yale University Press, pp. 453–514.

Cornwall, J. (1977) *Modern Capitalism: Its Growth and Transformation*, London: Martin Robertson.

Cyert, R. (1969) 'Uncertainty, behavioural rules, and the firms', *Economic Journal*, March.

Cyert, R. (1982) *Multinational Enterprise and Economic Analysis*, Cambridge: Cambridge University Press.

Cyert, R. and March, J.G. (1963) *A Behavioural Theory of the Firm*, Englewood Cliffs, NJ: Prentice-Hall.

Dahlman, C.J. and Westphal, L. (1983) 'The transfer of technology – issues in the acquisition of technological capability of developing countries', *Finance and Development* 20 (4).

Dasgupta, P. (1986) 'The theory of technological competition', in K. Binmore and P. Dasgupta (eds) *Economic Organisations as Games*, Oxford: Basil Blackwell.

Dasgupta, P. and Stiglitz, J. (1980a) 'Industrial structure and the nature of innovative activity', *Economic Journal* 90.

Dasgupta, P. and Stiglitz, J. (1980b) 'Uncertainty, industrial structure, and the speed of R&D', *Bell Journal of Economics* 11: 266–93.

David, P.A. (1975) 'The mechanisation of reaping in the ante-bellum midwest', in *Technical Choice, Innovation and Economic Growth*, Cambridge: Cambridge University Press, Chapter 4.

David, P.A. and van de Klundert, T. (1965) 'Biased efficiency growth and capital–labour substitution in the U.S., 1899–1960', *American Economic Review* 55: 357–94.

Davies, S. (1979) *The Diffusion of Process Innovations*, Cambridge: Cambridge University Press.

De Meza, D. and Dickinson, P.T. (1984) 'Risk preferences and the transaction costs', *Journal of Economic Organization and Behaviour*, June.

De Meza, D. and Klette, T. (1986) 'Is the market biased against risky R&D?' *Rand Journal of Economics* 17: 133–9.

Denison, E.F. (1967) *Why Growth Rates Differ: Postwar Experience in Nine Western Countries*, Washington, DC: Brookings Institution.

Denison, E.F. (1984) 'Accounting for slower economic growth: an update', in J.W. Kendrick (ed.) *International Comparisons of Productivity and Causes of the Slowdown*, Cambridge, MA: Ballinger.

Diamond, P., McFadden, D., and Rodriguez, M. (1965) 'Identification of the elasticity of substitution and the bias of technical change: an impossible theorem', Working Paper 62, University of California Institute for Business and Economic Research.

Diwan, R.K. (1970) 'About the growth path of firms', *American Economic Review* 60 (March).

Dixon, R. (1980) 'Hybrid corn revisited', *Econometrica* 48: 1451–62.

Dollar, D. (1986) 'Technological innovation, capital mobility, and the product cycle in north–south trade', *American Economic Review* 76: 177–90.

Domar, E.D. (1961a) 'On the measurement of technological change', *Economic Journal* (December): 709–29.

Domar, E.D. (1961b) 'The capital–output ratio in the United States: its variation and stability', in F.A. Lutz and D.C. Hague (eds) *The Theory of Capital*, Chapter 6.

Drandakis, E.M. and Phelps, E.S. (1966) 'A model of induced invention, growth and distribution', *Economic Journal* 76: 823–40.

Elster, J. (1983a), in *Explanation of Technical Change*, Cambridge: Cambridge University Press, Chapters 4, 6.

Elster, J. (1983b) *Sour Grapes: Studies in the Subversion of Rationality*, Cambridge: Cambridge University Press.

Enos, J. (1962) 'Invention and innovation in the petroleum refining industry', in *The Rate and Direction of Inventive Activity*, Princeton, NJ: Princeton University Press for the National Bureau of Economic Research.

Evenson, R. (1974) 'International diffusion of agrarian technology', *Journal of Economic History* 34: 51–73.

Evenson, R. (1984) in Z. Griliches (ed.) *R&D, Patents and Productivity*, Chicago, IL: University of Chicago Press for the National Bureau of Economic Research, Chapter 5.

Evenson, R. and Binswanger, H. (1978) 'Technology transfer and research resource allocation', in H. Binswanger and R. Ruttan (eds) *Induced Innovation: Technology, Institutions and Development*, Baltimore, MD: Johns Hopkins University Press, Chapter 6.

Evenson, R. and Kislev, Y. (1975) *Agricultural Research and Productivity*, New Haven; CT: Yale University Press.

Evenson, R. and Kislev, Y. (1976) 'A stochastic model of applied research', *Journal of Political Economy* 84: 265–81.

Fallenbuchl, Z. (1983) *East–West Technology Transfer: Poland*, Paris: OECD.

Fellner, W. (1971) 'Empirical support for the theory of induced innovations', *Quarterly Journal of Economics* 85: 580–604.

Ferguson, C.E. (1969) *The Neoclassical Theory of Production and Distribution*, Cambridge: Cambridge University Press.

Findley, R. (1978a) 'Relative backwardness, direct foreign investment, and the transfer of technology: a simple dynamic model', *Quarterly Journal of Economics* 92: 1–16.

Findley, R. (1978b) 'Some aspects of technology transfer and direct foreign investment', *American Economic Review* 68 (2): 275–9.

Fisher, F.M. and Temin, P. (1973) 'Returns of scale in research and development: what does the Schumpeterian hypothesis imply?', *Journal of Political Economy* 81: 56–70.

Fisher, F.M. and Temin, P. (1979) 'The Schumpeterian hypothesis: Reply', *Journal of Political Economy* 87: 386–9.

Flaherty, M.T. (1980a) 'Industry structure and cost-reducing investment', *Econometrica* 48: 1187–209.

Flaherty, M.T. (1980b) 'Timing patterns of new technology adoption in duopolostic industries', unpublished manuscript.

Forsund, F.R., Lovell, C.A.K., and Schmidt, P. (1980) 'A survey of frontier production functions and of their relationship to efficiency measurement', *Journal of Econometrics* 13 (1): 5–27.

Fransman, M. and King, K. (eds) (1984) *Technological Capability in the Third World*, London: Macmillan.

Freeman, C. (1982) *The Economics of Industrial Innovation*, 2nd edn, London: Frances Pinter (1st edn by Penguin, 1974).

Freeman, C., Clarke, J., and Soete, L. (1982) *Unemployment and Technical Innovation: A Study of Long Waves and Economic Development*, London: Frances Pinter.

Fudenberg, D. and Tirole, J. (1985) 'Preemption and rent equalization in the adoption of new technology', *Review of Economic Studies* 52: 383–401.

Fuss, M. and McFadden, D. (eds) (1978) *Production Economics: A Dual Approach to Theory and Applications*, Vols 1 and 2, Amsterdam: North-Holland.

Fuss, M., McFadden, D., and Mundlak, Y. (1978) 'A survey of functional forms in the economic analysis of production', in M. Fuss and D. McFadden (eds) *Production Economics: A Dual Approach to Theory and Applications*, Amsterdam: North-Holland, Vol. 1, pp. 217–68.

Futia, C.A. (1980) 'Schumpeterian competition', *Quarterly Journal of Economics* 94: 675–95.

Gabisch, G. (1975) 'A vintage capital model of international trade: the case of trade with second-hand machines', *Journal of International Economics* 5: 365–83.

Gerschenkron, A. (1962) *Economic Backwardness in Historical Perspective*, Cambridge, MA: Harvard University Press.

Ghatak, S. (1981) 'Technology transfer to developing countries: the case of the fertilizer industry', *Contemporary Studies in Economic and Financial Analysis* 27.

Gibbons, M. and Johnston, R. (1974) 'The roles of science in technological innovation', *Research Policy* 3: 220–42.

Gold, B. (ed.) (1977) *Research, Technological Change in Economic Analysis*, New York: Lexington Books.

Gold, B., Pierce, W.S., and Rosegger, G. (1970) 'Diffusion of major technological innovations in U.S. iron and steel manufacturing', *Journal of Industrial Economics* 18: 218–42.

Gold, B., *et al.* (1975) 'Diffusion of major technological innovations', in B. Gold (ed.) *Economics, Management and Environment*, Oxford: Pergamon.

Goldhar, J.D. (1974) 'Information, idea generation and technological innovation', in H.F. Davidson *et al.* (eds) *Technology Transfer*, Leiden: Noordhoff.

Gomulka, S. (1970) 'Extensions of the Golden Rule of Research of Phelps', *Review of Economic Studies*, January.

Gomulka, S. (1971) *Inventive Activity, Diffusion and the Stages of Economic Growth*, Aarhus: Institute of Economics and Aarhus University Press.

Gomulka, S. (1976a) 'Growth and the import of technology: Poland 1971–1980', *Cambridge Journal of Economics* 2 (1): 1–16.

Gomulka, S. (1976b) 'Do new factories embody best practice technology? New evidence', *Economic Journal* 86: 859–63.

Gomulka, S. (1977) 'Slowdown in Soviet industrial growth 1947–1975 reconsidered', *European Economic Review* 10: 37–49.

Gomulka, S. (1979) 'Britain's slow economic growth: low innovation versus increasing inefficiency', in W. Beckerman (ed.) *Slow Growth in Britain: Causes and Consequences*, Oxford: Oxford University Press.

Gomulka, S. (1986a) 'Soviet growth slowdown: duality, maturity and innovation', *American Economic Review, Papers and Proceedings*, May.

Gomulka, S. (1986b) *Growth, Innovation and Reform in Eastern Europe*, Brighton: Wheatsheaf.

Gomulka, S. (1987) 'Catching-up', *New Palgrave Dictionary of Economics*, London: Macmillan.

Gomulka, S. (1988) 'The Gerschenkron phenomenon and systemic factors in the post-1975 growth slowdown', *European Economic Review* 32 (1–2).

Gomulka, S. and Nove, A. (1985) *East–West Technology Transfer: Econometric Evaluation of the Contribution of West–East Technology Transfer to the East's Economic Growth*, Paris: OECD.

Gomulka, S. and Ostojic, S. (1986) 'Innovation activity in the Yugoslav economy', in S. Gomulka *Growth, Innovation and Reform in Eastern Europe*, Brighton: Wheatsheaf, pp. 62–73.

Gomulka, S. and Rostowski, J. (1988) 'An international comparison of material intensity', *Journal of Comparative Economics* (December): 475–501.

Gomulka, S. and Schaffer, M.E. (1987) 'Kaldor's stylized facts, and systemic and diffusion effects in productivity and growth', in G. Fink, G. Poll, and M. Rise (eds) *Economic Theory, Political Power and Social Justice*, New York: Springer-Verlag.

Gomulka, S. and Schaffer, M.E. (1989) 'A new method of long-run growth accounting with applications to the Soviet economy 1928–87 and the US Economy 1949–78', mimeograph.

Gomulka, S. and Sylwestrowicz, J.D. (1976) 'Import-led growth: theory and estimation', in F.L. Altman *et al.* (eds) *On the Measurement of Factor Productivities*, Göttingen: Vandenhoeck and Ruprecht.

Grabowski, H.G. (1968) 'The determinants of industrial R&D: a study of the chemical, drug and petroleum industries', *Journal of Political Economy*, March/April.

Grabowski, R. (1979) 'The implications of an induced innovation model', *Economic Development and Cultural Change* 27: 723–34.

Grant, J. (1979) 'Soviet machine tools: lagging technology and rising imports', in E.I. Hughes and J. Noren (eds) *Soviet Economy in a Time of Change*, Washington, D.C: US Government Printing Office, p. 555.

Greene, W.H. (1980a) 'Maximum likelihood estimation of econometric frontier functions', *Journal of Econometrics* 13 (1): 27–57.

Greene, W.H. (1980b) 'On the estimation of flexible frontier production model', *Journal of Econometrics* 13 (1): 101–17.

Gregory, R.G. and James, D.W. (1973) 'Do new factories embody best practice technology?', *Economic Journal* 83: 1135–55.

Griliches, Z. (1979) 'Issues in assessing the contribution of research and development of productivity growth', *Bell Journal of Economics* 10: 92–116.

Griliches, Z. (1980a) 'Hybrid corn revisited: a reply', *Econometrica* 48: 1463–5.

Griliches, Z. (1980b) 'R&D and the productivity slowdown', *American Economic Review* 70: 343–8.

Griliches, Z. (1980c) 'Returns to research and development expenditures in the private sector', in J.W. Kendrick and B.N. Vaccara (eds) *New Developments in Productivity Measurement and Analysis*, Chicago, IL: University of Chicago Press.

Griliches, Z. (ed.) (1984) *R&D, Patents, and Productivity*, Chicago, IL: University of Chicago Press for the National Bureau of Economic Research.

Griliches, Z. and Jorgenson, D.W. (1967) 'The explanation of productivity change', *Review of Economic Studies* 34: 249–83.

Griliches, Z. and Lichtenberg, F. (1984a) 'R&D and productivity at the industry level: is there still a relationship?', in Z. Griliches (ed.) *R&D, Patents, and Productivity*, Chicago, IL: University of Chicago Press.

Griliches, Z. and Lichtenberg, F. (1984b) 'Interindustry technology flows and productivity growth: a reexamination', *Review of Economics and Statistics* (May): 324–9.

Griliches, Z. and Mairesse, J. (1983) 'Comparing productivity growth; an exploration of French and U.S. industrial firm data', *European Economic Review* 21: 89–119.

Griliches, Z., Pakes, A., and Hall, B. (1987) 'The value of patents as indicators of inventive activity', in P. Dasgupta and P. Stoneman (eds) *Economic Policy and Technological Performance*, Cambridge: Cambridge University Press.

Grindley, P. (1986) 'A strategic analysis of the diffusion of innovations: theory and evidence', Ph. D. Thesis, London School of Economics.

Gruber, W.H., Metha, D., and Vernon, R. (1967) 'The R&D factor in international trade and international investment of U.S. industries', *Journal of Political Economy* 75: 20–37.

Haache, G. (1979) *The Theory of Economic Growth: An Introduction*, London: Macmillan.

Hahn, F.H. and Matthews, R.C.O. (1964) 'The theory of economic growth: a survey', *Economic Journal* 74: 779–902. (Reprinted in American Economic Association/Royal Economic Society, *Surveys of Economic Theory*, Vol. 2, London: Macmillan, 1965.)

Hall, R.E. (1968) 'Technical change and capital from the point of view of the dual', *Review of Economic Studies* 35 (1).

Hanson, P. (1981) *Trade and Technology in Soviet–Western Relations*, London: Macmillan.

Hanson, P. (1982) 'The end of import-led growth? Some observations on Soviet, Polish and Hungarian experience in the 1970s', *Journal of Comparative Economics* 6 (2): 130–47.

Hanson, P. and Pavitt, K. (1987) *The Comparative Economics of*

Research, Development and Innovation in East and West: A Survey, Harwood Academic Publishers.

Hébert, R.F. and Link, A.N. (1982) *The Entrepreneur: Mainstream Views and Radical Critiques*, New York: Praeger.

Heertje, A. (ed.) (1981) *Schumpeter's Vision: Capitalism, Socialism and Democracy After 40 Years*, New York: Praeger.

Helleiner, G.K. (1975) 'The role of multinational corporations in the less developed countries' trade in technology', *World Development*: 161–89.

Helpman, E. (1988) 'Growth, technical progress and trade', Working Paper 2592, National Bureau of Economic Research.

Hewett, A. (ed.) (1988) *Reforming the Soviet Economy: Equality versus Efficiency*, Washington, DC: Brookings Institution.

Hill, C.T. and Utterback, J.M. (eds) (1979) *Technological Innovation for a Dynamic Economy*, New York: Pergamon.

Hirschleifer, J. (1971) 'The private and social value of information and the reward to inventive activity', *American Economic Review* 61 (4): 561–74.

Hirschleifer, J. (1977) 'Economics from a biological viewpoint', *Journal of Law and Economics*, April.

Hirschleifer, J. (1978) 'Competition, cooperation and conflict in economics and biology', *American Economic Review* 68 (May).

Hirschleifer, J. and Rile, J.G. (1979) 'The analytics of uncertainty and information: an expository survey', *Journal of Economic Literature* 17 (December).

Hoffman, W.G. (1961), in F.A. Lutz and D.C. Hague (eds) *The Theory of Capital*, London: Institute of Economic Affairs and Macmillan.

Hong, W. and Krause, C.B. (1981) *Trade and Growth of the Advanced Developing Countries in the Pacific Basin: Papers and Proceedings of the Eleventh Pacific Trade and Development Conference*, Seoul: Korea Development Institute.

Hood, N. and Young, S. (1982) *The Economics of Multinational Enterprise*, Harlow, Essex: Longmans.

Horvat, B. (1974) 'Welfare of the common man in various countries', *World Development* 2 (7).

Hughes, E.I. and Noren, J. (1979) 'US and USSR: comparisons of GDP', in *Soviet Economy in a Time of Change*, Washington, DC: US Government Printing Office.

Hulten, C.R. (1975) 'Technical change and the reproducibility of capital', *American Economic Review* 65 (5): 956–65.

Hulten, C.R. (1978) 'Growth accounting with intermediate inputs', *Review of Economic Studies* 45 (3): 511–18.

Hymer, S. (1960) 'The international operation of national firms: a study of direct investment', Ph.D. Dissertation, Massachusetts Institute of Technology. (Published by MIT Press, Cambridge, MA, 1976.)

Hymer, S. (1976) *The International Operations of National Firms: A Study of Direct Foreign Investment*, Cambridge, MA: MIT Press.

Iwai, K. (1984a) 'Schumpeterian dynamics, Part I: An evolutionary model of innovation and imitation', *Journal of Economic Behaviour and Organization* 5: 159–90.

Iwai, K. (1984b) 'Schumpeterian dynamics, Part II: Technological progress, firm growth and "economic selection"', *Journal of Economic Behaviour and Organization* 5: 321–51.

Jaffe, A. (1986) 'Technological opportunity and spillovers of R&D', *American Economic Review* 76: 984–1001.

Jensen, R. (1982) 'Adoption and diffusion of an innovation of uncertain profitability', *Journal of Economic Theory* 27: 182–93.

Jensen, R. (1983) 'Innovation adoption and diffusion when there are competing innovations', *Journal of Economic Theory* 29: 161–71.

Jensen, R. and Thursby, M. (1987) 'A decisions theoretic model of innovation, technology transfer and trade', *Review of Economic Studies* 55: 631–48.

Jewkes, J., Sawers, D., and Stillerman, R. (1958) *The Sources of Invention*, London: Macmillan.

Jones, H.G. (1976) *An Introduction to Modern Theories of Economic Growth*, New York: McGraw-Hill.

Jones, R.W. (1970) 'The role of technology in the theory of international trade', in R. Vernon (ed.) *The Technology Factor in International Trade*, New York: Columbia University Press for the National Bureau of Economic Research, pp. 73–92.

Jorgenson, D.W. and Nishimizu, M. (1978) 'US and Japanese economic growth 1952–1974: an international comparison', *Economic Journal* 88: 702–26.

Kaldor, N. (1961) 'Capital accumulation and growth', in F.A. Lutz and D.C. Hague (eds) *The Theory of Capital*, London: Institute of Economic Affairs and Macmillan.

Kalecki, M. (1954) *Theory of Economic Dynamics*, London: Allen & Unwin.

Kalecki, M. (1962) 'Observations on the theory of growth', *Economic Journal* 72 (March).

Kamien, M.I. and Schwartz, N.L. (1975) 'Market structure and innovation: a survey, *Journal of Economic Literature* 13: 1–37.

Kamien, M.I. and Schwartz, N.L. (1976) On the degree of rivalry for maximum innovative activity', *Quarterly Journal of Economics* 90: 245–60.

Kamien, M.I. and Schwartz, N.L. (1982) *Market Structure and Innovation*, Cambridge: Cambridge University Press.

Katsuhito, I. (1984) 'Schumpeterian dynamics: an evolutionary model of innovation and imitation', *Journal of Economic Behaviour and Organization*, June.

Kendrick, J.W. (1977) *Understanding Productivity: An Introduction to the Dynamics of Productivity Change*, Baltimore, MD: Johns Hopkins University Press.

Kendrick, J.W. (1980) 'Total investment, capital, and economic growth', in R.C.O. Matthews (ed.) *Economic Growth and Resources*, Vol. 2, *Trends and Factors*, London: Macmillan, Chapter 5.

Bibliography

Kendrick, J.W. and Grossman, E.S. (1980) *Productivity in the United States: Trends and Cycles*, Baltimore, MD: Johns Hopkins University Press.

Kennedy, C. (1964) 'Induced bias in innovation and the theory of distribution', *Economic Journal*, September.

Kennedy, C. (1973) 'A generalisation of the theory of induced bias in technical progress', *Economic Journal* 83.

Kennedy, C. and Thirlwall, A.P. (1972) 'Technical progress: a survey', *Economic Journal*, March.

Kindleberger, C.P. (1969) *American Business Abroad*, New Haven, CT: Yale University Press.

Klette, T. and de Meza, D. (1986) 'Is the market biased against risky R&D?' *Rand Journal of Economics* 17: 133–9.

Kojima, K. (1977) *The Theory on Direct Investment*, Tokyo: Diamond.

Koniunturnyi Instytut (1926) *Mirovoe Khoziaistvo Statisticheskii Sbornik za 1913–25 gg*, Moscow.

Kornai, J. (1971) *Anti-Equilibrium*, Amsterdam: North-Holland.

Kornai, J. (1980) *The Economics of Shortage*, Amsterdam: North-Holland.

Kornai, J. (1985) 'Gomulka on the soft budget constraint: a reply', *Economy of Planning* 19 (2).

Kravis, I.B. (1976) 'International comparisons of productivity', *Economic Journal* 86: 1–44.

Kravis, I.B. and Lipsey, R. (1989) 'The effect of multinational firms' foreign operations on their domestic employment', *Working Paper 2760*, National Bureau of Economic Research.

Krugman, P. (1974) 'A model of innovations, technology transfer, and the world distribution of income', *Journal of Political Economy* 87: 253–66.

Kubielas, S. (1980) 'Mechanism of technological progress under the financial/economic system in Polish industry', mimeograph, Finance Institute, Warsaw.

Kuznets, S. (1971) *Economic Growth of Nations*, Cambridge, MA: Harvard University Press.

Kuznets, S. (1973) 'Modern economic growth: findings and reflections', *American Economic Review* 63 (June).

Lall, S. (1982) *Developing Countries as Exporters of Technology: A First Look at the Indian Experience*, London: Macmillan.

Landes, D.S. (1980) *The Unbound Prometheus: Technological Change and Industrial Development in Western Europe from 1750 to the Present*, 2nd edn, Cambridge: Cambridge University Press.

Lee, T. and Wilde, L. (1980) 'Market structure and innovation: a reformulation', *Quarterly Journal of Economics* 194: 429–36.

Leibenstein, H. (1966) 'Allocative efficiency of *X*-efficiency', *American Economic Review* 56. (Updated version: 'Aspects of the *X*-efficiency theory of the firm', *Bell Journal of Economics* 6 (2), Autumn 1975; also 'The missing link' – micro-micro theory?', *Journal of Economic Literature* (June): 477–502.)

Leonard, W.N. (1971) 'Research and development in industrial growth', *Journal of Political Economy* 79 (2): 232–56.

Leontief, W.W. (1947) 'Introduction to a theory of the internal structure of functional relationships', *Econometrica* 15: 361–73.

Levin, R.C. (1978) 'Technical change, barriers to entry and market structure', *Econometrica* 45: 347–61.

Levin, R. and Reiss, P. (1984) 'Tests of a Schumpeterian model of R&D and market structure', in Z. Griliches (ed.) *R&D, Patents, and Productivity*, Chicago, IL: University of Chicago Press, pp. 175–204. (See also the comments by P. Tandon, pp. 205–8.)

Levy, D.M. and Terleckyj, N.E. (1983) 'Effects of government R&D on private R&D investment and productivity: a macroeconomic analysis', *Bell Journal of Economics* 14: 551–61.

Link, A.N. (1980) 'Firm size and efficient entrepreneurial activity: a reformulation of the Schumpeter hypothesis', *Journal of Political Economy* 88: 771–82.

Link, A.N. (1981) *Research and Development in U.S. Manufacturing*, New York: Praeger.

Link, A.N. (1987) *Technological Change and Productivity Growth*, New York, Harwood Academic Publishers.

Lotka, A.J. (1926) 'The frequency distribution of scientific productivity', *Journal of the Washington Academy of Sciences* 16: 317.

Loury, G.C. (1979) 'Market structure and innovation', *Quarterly Journal of Economics* 93: 395–410.

Lutz, F.A. and Hague, D.C. (eds) (1961) *The Theory of Capital*, London: Institute of Economic Affairs and Macmillan.

McAuley, A. (1985) 'Central planning, market socialism and rapid innovation', in M. Schaffer (ed.) *Technology Transfer and East–West Relations*, London: Croom Helm.

McCain, R.A. (1972) 'Induced technical progress and the price of capital goods', *Economic Journal* 82: 327.

McCain, R.A. (1974) 'Induced bias in technical innovation including product innovation in a model of economic growth', *Economic Journal* 84: 959–66.

McFadden, D. (1978), in M. Fuss and D. McFadden (eds) *Production Economics: A Dual Approach to Theory and Applications*, Amsterdam: North Holland, Vol. 1, Chapter IV.1.

Machlup, F. (1962) *The Production and Distribution of Knowledge in the US*, Princeton, NJ: Princeton University Press.

Machlup, F. (1967) 'Theories of the firm: marginalist behavioural managerial', *American Economic Review*, March.

Machlup, F. (1980) *Knowledge: Its Creation, Distribution, and Economic Significance*, Vol. 1, *Knowledge and Knowledge Production*, Princeton, NJ: Princeton University Press.

Maddison, A. (1979) 'Long run dynamics of productivity growth', *Banca Nazionale del Lavoro Quarterly Review*, March. (Most data from this paper appeared also in Beckerman (1979: Chapter 10) and Matthews (1980: Chapter 1).)

Bibliography

Maddison, A. (1982) *Phases of Capitalist Development*, Oxford: Oxford University Press.

Maddison, A. (1984) 'Comparative analysis of the productivity situation in the advanced capitalist countries', in J.W. Kendrick (ed.) *International Comparisons of Productivity and Causes of the Slowdown*, Cambridge, MA: Ballinger.

Maddison, A. (1987) 'Growth and slowdown in advanced capitalist economies: techniques of quantitative assessment', *Journal of Economic Literature* 25: 649–98.

Mansfield, E. (1961) 'Technical change and the rate of imitation', *Econometrica* 2: 741–66.

Mansfield, E. (1963) 'Intrafirm rates of diffusion of an innovation', *Review of Economics and Statistics*, 45: 348–59.

Mansfield, E. (1968) *Industrial Research and Technological Innovation*, New York: Norton.

Mansfield, E. (1980) 'Basic research and productivity increase in manufacturing', *American Economic Review* 70: 863–73.

Mansfield, E. (1983) 'Technological change and market structure: an empirical study', *American Economic Review* 73: 205–9.

Mansfield, E., Rapoport, J., Romeo, A., Wagner, S., and Minasian, J. (1969) 'Research and development, production functions and rates of return', *American Economic Review* 59 (May).

Mansfield, E., Rapoport, J., Schnee, J., Wagner, S., and Hamburger, M. (1971) *Research and Innovation in the Modern Corporation*, New York: Norton.

Mansfield, E., Rapoport, J., Romeo, A., Villani, E., Wagner, S., and Husic, F. (1977a) *The Production and Application of New Industrial Technologies*, New York: Norton.

Mansfield, E., Rapoport, J., Romeo, A., Wagner, S., and Beardsley, G. (1977b) 'Social and private rates of return from industrial innovations', *Quarterly Journal of Economics* 91: 221–40.

Mansfield, E., Schwartz, M., and Wagner, S. (1981) 'Imitation costs and patents: an empirical study', *Economic Journal* 91: 907–18.

Mansfield, E., Romeo, A., Schwartz, M., Teece, D., Wagner, S., and Brach, P. (1982) *Technology Transfer, Productivity, and Economic Policy*, New York: Norton.

Marer, P. (1986) *East–West Technology Transfer: Czechoslovakia*, Paris: OECD.

Martens, J.A. and Young, J.P. (1979) 'Soviet implementation of domestic inventions – first results', in *Soviet Economy in a Time of Change*, Washington, DC: US Government Printing Office, Vol. 1, pp. 472–510.

Martin, F. (1986) *Technology and Economic Development*.

Matthews, R.C.O. (1973) 'The contribution of science and technology to economic development', in B. Williams (ed.) *Science and Technology in Economic Growth*, London: Macmillan.

Matthews, R.C.O. (ed.) (1980) *Economic Growth and Resources*, Vol. 2, *Trends and Factors*, London: Macmillan.

Matthews, R.C.O. (1984) 'Darwinism and economic change', *Oxford Economic Papers, Supplement*, November.

Melvin, J.R. (1969) 'Intermediate goods and technical change', *Economica* 36: 401–7.

Mensch, G. (1974) 'Institutional barriers to the science and technology interaction', H.F. Davidson *et al.* (eds) *Technology Transfer*, Leyden: Nordhoff.

Mensch, G. (1978) '1984: A new push for basic innovation?', *Research Policy* 7 (2): 108–22.

Minasian, J.R. (1962) 'The economies of research and development', in R. Nelson (ed.) *Rate and Direction of Inventive Activity*, Princeton, NJ: Princeton University Press.

Morishima, M. (1964) *Equilibrium, Stability and Growth*, Oxford: Oxford University Press.

Morishima, M. (1968) 'Economic growth: mathematical theory', *International Encyclopedia of the Social Sciences* 4: 417–22.

Morishima, M. (1976) *The Economic Theory of Modern Society*, Cambridge: Cambridge University Press.

Morishima, M. and Saito, M. (1968) 'An economic test of Sir John Hicks' theory of biased induced investigations', in J. Wolfe (ed.) *Value, Capital and Growth: Papers in Honour of Sir John Hicks*, Chicago, IL: Aldine.

Mowery, D.C. and Rosenberg, N. (1979) 'The influence of market demand upon innovation: a critical review of some recent empirical studies', *Research Policy* 8: 103–53. (Reprinted in N. Rosenberg, *Inside the Black Box: Technology and Economics*, New York: Cambridge University Press, 1982, Chapter 10, pp. 193–241.)

Nadiri, M.I. (1970) 'Some approaches to the theory and measurement of total factor productivity: a survey', *Journal of Economic Literature* 8 (4): 1137–77.

Nadiri, M.I. (1972) 'International studies of factor inputs and total factor productivity: a brief survey', *Review of Income and Wealth* (June): 129–54.

Nasbeth, L. and Ray, G.F. (1974) *Diffusion of New Industrial Processes*, Cambridge: Cambridge University Press.

Nelson, R.R. (1968) 'A diffusion model of international productivity differences in manufacturing industry', *American Economic Review* 58: 1219–47.

Nelson, R.R. (1980) 'Production sets, technological knowledge and R&D: fragile and overworked constructions for analysis of productivity growth?' *American Economic Review*, May.

Nelson, R.R. (1981) 'Research on productivity growth and productivity differences: dead ends and new departures', *Journal of Economic Literature* 19: 1029–64.

Nelson, R. and Norman, D. (1977) 'Technological change and factor-mix over the product cycle: a model of dynamic corporation advantage', *Journal of Development Economics* 4: 3–24.

Nelson, R.R. and Phelps E.S. (1966) 'Investment in humans, technological diffusion and economic growth', *American Economic Review* 56: 66–75.

Nelson, R.R. and Winter, S.G. (1974) 'Neoclassical vs. evolutionary theories of economic growth: critique and prospectus', *Economic Journal* 84: 886–905.

Nelson, R.R. and Winter, S.G. (1977a) 'In search of a useful theory of innovation', *Research Policy* 6: 36–76.

Nelson, R.R. and Winter, S.G. (1977b) 'Simulation of Schumpeterian competition', *American Economic Review*, May.

Nelson, R.R. and Winter, S.G. (1978) 'Forces generating and limiting concentration under Schumpeterian competition', *Bell Journal of Economics* 9: 524–48.

Nelson, R.R. and Winter, S.G. (1982a) 'The Schumpeterian tradeoff revisited', *American Economic Review* 72: 114–32.

Nelson, R.R. and Winter, S.G. (1982b) *An Evolutionary Theory of Economic Change*, Cambridge, MA: Belknap Press.

Nelson, R.R., Peck, M.J., and Kalachek, E.D. (1967) *Technology, Economic Growth, and Public Policy*, Washington, DC: Brookings Institution.

Nelson, R.R., Winter, S.G., and Schuette, H.L. (1976) 'Technical change in an evolutionary model', *Quarterly Journal of Economics* 40: 90–118.

Nishimizu, M. and Page, J.M. (1981) *Regional and Sectoral Productivity Performance of Yugoslavia, 1965–1978*, Washington, DC: World Bank Development Economics Department.

Nolting, L.E. and Feshbach, M. (1979) 'R&D employment in the USSR: definitions, statistics and comparison, in United States Congress, Joint Economic Committee', in *Soviet Economy in a Time of Change*, Washington, DC: US Government Printing Office.

Nordhaus, W.D. (1967) 'The optimal rate and direction of technical change', in K. Shell (ed.) *Essays in the Theory of Optimal Economic Growth*, Cambridge, MA: MIT Press.

Nordhaus, W.D. (1969a) 'An economic theory of technological change', *American Economic Review, Papers and Proceedings* (May): 18–28.

Nordhaus, W.D. (1969b) *Invention, Growth and Welfare*, Cambridge, MA: MIT Press.

Nordhaus, W.D. (1973) 'Some skeptical thoughts on the theory of induced innovation', *Quarterly Journal of Economics* 87: 209–19.

Norris, K. and Vaizey, J. (1973) *The Economics of Research and Technology*, London: Allen & Unwin.

North, D.C. and Thomas, R.P. (1970) 'An economic theory of the growth of the Western world', *Economic History Review* 22: 1–17.

North, D.C. and Thomas, R.P. (1973) *The Rise of the Western World*, Cambridge: Cambridge University Press.

OECD (1986) *R and D, Invention and Competitiveness*, Paris: OECD.

OECD (1989) 'DSTI: Technology and economic growth in a vintage model framework', mimeograph.

Olson, M. (1982) *The Rise and Decline of Nations*, New Haven, CT: Yale University Press.

Oman, C. (1984) *New Forms of International Investment in Developing Countries*, Paris: OECD Development Centre.

Onida, F. and Viesti, G. (eds) (1987) *Italian Multinationals*, London: Croom Helm.

Oster, S. (1982) 'The diffusion of innovation among steel firms: the basic oxygen process', *Bell Journal of Economics* 13: 45–56.

Ozawa, T. (1984) *New Forms of Investment by Japanese Firms*, Paris: OECD Development Centre.

Pakes, A. (1985) 'On patents, R&D and the stock market rate of return', *Journal of Political Economy* 93: 390–409.

Pakes, A. and Griliches, Z. (1984) 'Patents and R&D at the firm level: a first look', in Z. Griliches (ed.) *R&D, Patents, and Productivity*, Chicago, IL: University of Chicago Press, pp. 55–72.

Pakes, A. and Schankerman, M. (1984) 'An exploration into the determinants of research intensity', in Z. Griliches (ed.) *R&D, Patents and Productivity*, Chicago, IL: University of Chicago Press, pp. 209–32.

Pasinetti, L. (1981) *Structural Change and Economic Growth*, Cambridge: Cambridge University Press.

Pavitt, K. and Soete, L. (1982) 'International differences in economic growth and the international location of innovation', in H. Giersch (ed.) *Emerging Technologies: Consequences for Economic Growth, Structural Change and Employment*, JCB Mohr.

Pavitt, K., Robson, M., and Townsend, J. (1987) 'The size distribution of innovating firms in the UK: 1945–83', *Journal of Industrial Economics*, May.

Peterson, W. (1979) 'Total factor productivity in the UK: a disaggregated analysis', in K.D. Patterson and K. Schott (eds) *The Measurement of Capital: Theory and Practice*, London: Macmillan, Chapter 8.

Phelps, E. (1966) 'Models of technical progress and the Golden Rule of research', *Review of Economic Studies* 33: 133–45.

Phillips, A. (1971) *Technology and Market Structure*, Lexington, MA: D.C. Heath.

Poznanski, K. (1980) 'A study of technical innovation in Polish industry', *Research Policy* 9: 232.

Poznanski, K. (1986) 'Competition between Eastern Europe and developing countries in the Western market for manufactured goods', in *East European Economics: Slow Growth in the 1980s*, Vol. 2, Washington, DC: US Government Printing Office.

Poznanski, K. (1987) *Technology, Competition and the Soviet Bloc in the World Market*, Berkeley, CA; IIS.

Price, D.J. de S. (1963) *Little Science, Big Science*, New York: Columbia University Press.

Quirmbach, J. (1986) 'The diffusion of new technology and the market for an innovation', *Rand Journal of Economics* 17 (1): 33–47.

Ray, G.F. (1984) *The Diffusion of Mature Technologies*, Cambridge: Cambridge University Press.

Bibliography

Ray, G.F., and Uhlmann, L. (1979) *The Innovation Process in the Energy Industries*, Cambridge: Cambridge University Press.

Reinganum, J.F. (1981a) 'Dynamic games of innovation', *Journal of Economic Theory* 25: 21–41.

Reinganum, J.F. (1981b) 'Market structure and the diffusion of new technology', *Bell Journal of Economics* 12: 618–24.

Reinganum, J.F. (1981c) 'On the diffusion of new technology: a game theoretic approach', *Review of Economic Studies* 48: 395–405.

Reinganum, J.F. (1983) 'Technology adoption under imperfect information', *Bell Journal of Economics* (Spring): 57–63.

Reinganum, J.F. (1985) 'Innovation and industry evolution', *Quarterly Journal of Economics* 100: 81–99.

Reynolds, W. (1983) 'The spread of economic growth to the Third World: 1850–1980', *Journal of Economic Literature* 21: 941–80.

Rodriguez, C.A. (1979) 'A comment on Fisher and Temin on the Schumpeterian hypothesis', *Journal of Political Economy* 87: 383–5.

Rogers, E.M. (1983) *Diffusion of Innovations*, 3rd edn, New York: Macmillan.

Romeo, A.A. (1975) 'Interindustry and interfirm differences in the rate of diffusion of an innovation', *Review of Economics and Statistics* 57: 311–19.

Romeo, A.A. (1977) 'The rate of imitation of a capital-embodied process innovation', *Econometrica* 44: 63–9.

Romer, P.M. (1986) 'Increasing returns and long-run growth', *Journal of Political Economy* 94: 1002–37.

Rosenberg, N. (1969) 'The direction of technological change: inducement mechanisms and focusing devices', *Economic Development and Cultural Change* 18: 1–24.

Rosenberg, N. (ed.) (1971) *The Economics of Technological Change*, Harmondsworth: Penguin.

Rosenberg, N. (1974) 'Science, invention and economic growth', *Economic Journal* 100: 725–9.

Rosenberg, N. (1976) *Perspective on Technology*, New York: Cambridge University Press.

Rosenberg, N. (1982) *Inside the Black Box: Technology and Economics*, 2nd edn, Cambridge: Cambridge University Press.

Rosenberg, N. and Birdzell, L.E. (1986) *How the West Grew Rich*, London: I.B. Tauris.

Rostow, W.W. (1960) *The Stages of Economic Growth: A Non-Communist Manifesto*, Cambridge: Cambridge University Press.

Rostow, W.W. (1978) *The World Economy: History and Prospects*, Austin, TX: University of Texas Press.

Rothschild, M. (1971) 'On the cost of adjustment', *Quarterly Journal of Economics* 85: 605–22.

Ruttan, V.W. (1959) 'Usher and Schumpeter on invention, innovation and technological change', *Quarterly Journal of Economics* 73: 596–606.

Ruttan, V.W. (1977) 'The Green Revolution: seven generalizations', *International Development Review* 19: 16–23.

Ruttan, V.W. and Hayami, Y. 'Technology transfer and agricultural developments', *Technology and Culture* 14: 119–51.

Ruttan, V.W., Binswanger, H., Hayami, Y., Wade, W., and Weber, A. (1978) 'Factor productivity and growth: a historical interpretation', in H. Binswanger and V.W. Ruttan (eds) *Induced Innovation: Technology, Institutions and Development*, Baltimore, MD: Johns Hopkins University Press.

Ryan, B. and Gross, N. (1943) 'The diffusion of hybrid seed corn in two Iowa communities', *Rural Sociology* 7 (March): 15–24.

Rymes, T.K. (1971) *On Concepts of Capital and Technical Change*, Cambridge: Cambridge University Press.

Rymes, T.K. (1973) 'The measurement of capital and total factor productivity in the context of the Cambridge theory of capital', *Review of Income and Wealth* 18 (1): 79–108.

Rymes, T.K. (1983) 'More on the measurement of total factor productivity', *Review of Income and Wealth* 29: 297–316.

Sahal, D. (ed.) (1982) *The Transfer and Utilization of Technical Knowledge*, Lexington, MA: D.C. Heath.

Salter, W.E.G. (1966) *Productivity and Technical Change.* 2nd edn, Cambridge: Cambridge University Press.

Samuelson, P. (1965) 'A theory of induced innovation along Kennedy–Wiesacker lines', *Review of Economics and Statistics* 47: 343–56.

Samuelson, P. (1966) 'Rejoinder: Agreements, disagreements, doubts, and the case of induced Harrod-neutral technical change', *Review of Economics and Statistics* 48: 444–8.

Sato, R. (1970) 'The estimation of biased technical progress and the production function', *International Economic Review* 11: 179–208.

Sato, R. and Suzawa, G.S. (1983) *Research and Productivity: Endogenous Change*, Boston, MA: Auburn House.

Sato, R. and Tsutsui, S. (1984) 'Technical progress, the Schumpeterian hypothesis, and market structure', *Zeitschrift für Nationalokonomie (Supplement)* 44: 1–37.

Schaffer, M. (1989) 'Redistribution of profit, financial flows and economic reform in Polish industry: evidence from the "Lista 500"', mimeograph London School of Economics.

Schankerman, M. and Pakes, A. (1984), in Z. Griliches (ed.), *R&D, Patents, and Productivity*, Chicago, IL: University of Chicago Press.

Schankerman, M. and Pakes, A. (1986) 'Estimates of the value of patent rights in European countries during the post-1950 period', *Economic Journal* 96: 1052–76.

Scherer, F.M. (1967) 'Research and development resource allocation under rivalry', *Quarterly Journal of Economics* 81: 359–94.

Scherer, F.M. (1980) *Industrial Market Structure and Economic Performance*, 2nd edn, Chicago, IL: Rand McNally.

Scherer, F.M. (1982a) 'Demand-pull and technological invention: Schmookler revisited', *Journal of Industrial Economics* 30: 225–37.

Scherer, F.M. (1982b) 'Inter-industry technology flows and productivity growth', *Review of Economics and Statistics* 6: 627–34.

Scherer, F.M. (1983) 'R&D and declining productivity growth', *American Economic Review* 73: 215–18.

Scherer, F.M. (1984) *Innovation and Growth: Schumpeterian Perspectives*, Cambridge, MA: MIT Press.

Schmidt, P. (1976) 'On the statistical estimation of parametric frontier production functions,' *Review of Economics and Statistics* 58 (2): 238–9.

Schmookler, J. (1966) *Invention and Economic Growth*, Cambridge, MA: Harvard University Press.

Schmookler, J. (1972) *Patents, Invention and Economics Growth: Data and Selected Essays*, Cambridge, MA: Harvard University Press.

Schott, K. (1977) 'Investment in private industrial research and development in Britain', *Journal of Industrial Economics* 81–99.

Schumpeter, J.A. (1928) 'The instability of capitalism', *Economic Journal* 38 (151): 386.

Schumpeter, J.A. (1934) *The Theory of Economic Development*, Cambridge, MA: Harvard University Press. (Reprinted 1974, Oxford University Press, Oxford.)

Schumpeter, J.A. (1939) *Business Cycles: A Theoretical, Historical and Statistical Analysis of the Capitalist Process*, New York: McGraw Hill. (Abridged version: R. Fels, New York, 1964.)

Schumpeter, J.A. (1942) *Capitalism, Socialism, and Democracy*, New York: Harper.

Shah, A. and Desai, M. (1981) 'Growth cycles with induced technical change', *Economic Journal* 91: 1006–10.

Shen, T.Y. (1970) 'Economics of scale, Penrose effect, growth of plants and their size distribution', *Journal of Political Economy* 78 (4).

Simon, H.A. (1951) 'Effects of technological change in a Leontief model', in T.C. Koopmans (ed.) *Activity Analysis of Production and Allocation*, New York: Wiley, pp. 260–81.

Simon, H.A. (1957) *Models of Man, Social and National*, New York: Wiley.

Simon, H.A. (1962) 'New developments in the theory of the firm', *American Economic Review* 52 (May).

Simon, H.A. (1983) *Reason in Human Affairs*, Oxford: Basil Blackwell.

Slama, J. (1983) 'Gravity model and its estimations for international flows of engineering products, chemicals and patent applications', *Acta Oeconomica* 30 (2): 241–53.

Slama, J. (1986) 'An international comparison of sulphur dioxide emissions', *Journal of Comparative Economics* 10 (3): 277–92.

Smith, M.A.M. (1974) 'International trade in second-hand machines', *Journal of Development Economics* 1: 261–78.

Smith, M.A.M. (1976) 'Trade, growth and consumption in alternative models of capital accumulation', *Journal of International Economics* 6: 371–84.

Soete, L. and Turner, R. (1984) 'Technology diffusion and the rate of technical change', *Economic Journal* 94: 612–23.

Solow, R.M. (1956) 'A contribution to the theory of economic growth', *Quarterly Journal of Economics* 70: 65–94.

Solow, R.M. (1957) 'Technical change and the aggregate production function', *Review of Economics and Statistics* 39: 312–20.

Solow, R.M. (1962) 'Technical progress, capital formation and economic growth', *American Economic Review* 52: 76–86.

Solow, R.M. (1970) *Growth Theory: An Exposition,* New York: Oxford University Press.

Solow, R.M. and Temin, P. (1985) 'The inputs for growth', in J. Mokyr (ed.) *The Economics of the Industrial Revolution,* London: Allen & Unwin, Chapter 3.

Spence, M. (1984) 'Cost reduction, competition, and industry performance', *Econometrica* 52: 101–21.

Steedman, I. (1985) 'On the "impossibility" of Hicks-neutral technical change', *Economic Journal* 95: 746–58.

Stevenson, R. (1980) 'Measuring technological bias', *American Economic Review* 70: 162–73.

Stewart, F. and James, J. (1982) *The Economics of New Technology in the Developing Countries,* London: Frances Pinter.

Stoneman, P. (1976) *Technology Diffusion and the Computer Revolution: The UK Experience,* Cambridge: Cambridge University Press.

Stoneman, P. (1981) 'Infra-firm diffusion Bayesian learning and profitability', *Economic Journal* 91: 375–88.

Stoneman, P. (1983) *The Economic Analysis of Technological Change,* London: Oxford University Press.

Stoneman, P. and David, P.A. (1986) 'Adoption subsidies vs. information provision as instruments of technology policy', *Economic Journal (Supplement)* 96: 142S–150S.

Stoneman, P. and Ireland, N.J. (1983) 'The role of supply factors in the diffusion of new process technology', *Economic Journal (Supplement)* 93: 65S–77S.

Stressler, E. (1980) 'Models of investment-dependent economic growth revisited', in R.C.O. Matthews (ed.) *Economic Growth and Resources,* Vol. 2, *Trends and Factors,* London: Macmillan, Chapter 8.

Summers, R. and Heston, A. (1984) 'Improved international comparisons of real product and its composition 1950–1980', *Review of Income and Wealth* 30 (2), June.

Svejnar, J. and Smith, S.C. (1982) 'The economics of joint ventures in centrally planned and labour-managed economies', *Journal of Comparative Economics* 6: 148–72.

Svejnar, J. and Smith, S.C. (1984) 'The economics of joint ventures in less developed countries', *Quarterly Journal of Economics* (February): 149–67.

Swan, P.L. (1973) 'The international diffusion of an innovation', *Journal of Industrial Economics* 22 (1): 61–70.

Takayama, A. (1974) 'On biased technological progress', *American Economic Review* 64 (4): 631–9.

Takayama, A. (1977) 'Technology transfer by multinational firms: the resource cost of transferring technological know-how', *Economic Journal* 87: 242–61.

Teece, D.J. (1976) *The Multinational Corporation and the Resource Cost of International Technology Transfer*, Cambridge, MA: Ballinger.

Teece, D.J. (1977) 'Technology transfer by multinational firms: the resource cost of transferring technological knowhow', *Economic Journal* 87: 242–61.

Teitel, S. (1984) 'Technology creation in semi-industrial economies', *Journal of Development Economics* 6 (1/2): 39–61.

Terlecky N.E. (1980) 'Direct and indirect effects of industrial research and development in the productivity growth of industries', in J.W. Kendrick and B.N. Vaccara (eds) *New Developments in Productivity Measurement and Analysis*, Chicago, IL: University of Chicago Press.

Terlecky, N.B. (1982) 'R&D and U.S. industrial productivity in the 1970s', in D. Sahal (ed.) *The Transfer and Utilization of Technical Knowledge*, Lexington, MA: D.C. Heath.

Thirtle, C.G. and Ruttan, V.W. (1987) *The Role of Demand and Supply in the Generation and Diffusion of Technical Change.* New York: Harwood Academic Publishers.

Usher, D. (1980) *The Measurement of Economic Growth*, Oxford: Basil Blackwell.

Uzawa, H. (1965) 'Optimal technical change in an aggregative model of economic growth, *International Economic Review* 6: 18–31.

Veblen, T. (1915) *Imperial Germany and the Industrial Revolution*, London: Macmillan.

Vernon, R. (1966) 'International investment and international trade in the product cycle', *Quarterly Journal of Economics* 80: 190–207.

Vernon, R. (1979) 'The product cycle hypothesis in a new international environment', *Oxford Bulletin of Economics and Statistics* 41: 255–67.

Wan, H.Y. (1971) *Economic Growth*, New York: Harcourt Brace Jovanovich.

Weitzman, M. (1970) 'Soviet post-war economic growth and capital–labour substitution', *American Economic Review* 60 (September).

von Weizsacker, C.C. (1966) 'Tentative notes on a two sector model with induced technical progress', *Review of Economic Studies* 33: 245–51.

von Weizsacker, C.C. (1980) *Barriers to Entry: A Theoretical Treatment*, Berlin: Springer-Verlag.

Whitaker, J.K. (1970) 'Harrod-neutral technical progress and the possibility of steady growth', *Revista Internationale di Scienze Economiche e Commerciali* 2.

Williams, B.R. (1973) *Science and Technology in Economic Growth*, London: Macmillan.

Williamson, J.G. (1976) 'Technology, growth and history', *Journal of Political Economy, Part 1* 8: 809–20.

Williamson, D.E. (1975) *Markets and Hierarchies: Analysis and Antitrust Implications. A Study in the Economies of Internal Organization*, New York: Free Press.

Winter, S.G. (1971) 'Satisficing, selection and innovative remnant', *Quarterly Journal of Economics*, May.

Winter, S.G. (1984) 'Schumpeterian competition in alternative technological regimes', *Journal of Economic Behaviour and Organisation* 5: 287–320.

Wyatt, G. (1986) *The Economics of Invention: A Study of the Determinants of Inventive Activity*, Brighton: Harvester Press.

Wyatt, S. (1985) 'Patents and multinational corporations', *World Patent Information* 7 (3).

You, J.K. (1976) 'Embodied and disembodied technical progress in the United States, 1929–1968', *Review of Economics and Statistics* 123–7.

Zaleski, E. and Wiener, H. (1980) *East–West Technology Transfer: Grand Analysis*, Paris: OECD.

Zaleski, E. *et al.* (1969) *Science Policy in the USSR*, Paris: OECD.

Index

Index

process intensity 7
product cycle 15–16; international
    199–203; theory 83
product innovation *see* innovation,
    product
production possibility frontier
    (production function) 8–9, 67,
    77
production possibility set 8–9
production processes (activities)
    4–5
production sector, defined 3–4
productivity, of capital 221–2; of
    labour 152–5, 190–2, 195;
    primary sector 19
productivity gap 152–5
productivity growth, long-term 223;
    variations across countries
    151–2
progress goods 3
public goods, invention as 25–6

Quirmbach, J. 92

R&D (research and development),
    by firm size 43–4, 49; choice of
    projects 27–9; concentration 43;
    growth of sector 34–5; intensity
    41–3, 49, 76; interfirm
    expenditure variations 45–7;
    international distribution 192–6;
    in mature industries 16–17;
    over-fishing model 26; private
    rate of return 40–1; social rate of
    return 39–40; structure 11–13;
    of suppliers 14; world expansion
    177–8
Ramsey, F.P. 19
Ray, G.F. 89
Reinganum, J.F. 92
relative technology gap 160–1
reproducibility problem 133
research intensity 54; optimum 171
Riess, P. 49
Rile, J.G. 26
Robinson, J. 122
Rodriquez, M. 127–8
Romeo, A.A. 89

Rosenberg, N. 28, 44–5
Rostow, W.W. 151, 156
Rostowski, J. 97
Ruttan, V.W. 80
Ryan, B. 80
Rymes, T.K. 131, 136, 139–41

S-shaped: curve diffusion pattern
    93
sales growth, and R&D intensity 41,
    43
Salter, W.E.G. 119
Samuelson, P.A. 142
satisficers 74
Sato, R. 128
scale 7–8, 16, 26, 195, *see also*
    economies of scale
scale effects 29
Schaffer, M. 113, 162–4
Schankerman, M. 41
Scherer, F.M. 49
Schmookler, J. 12, 31
Schott, K. 41
Schumpeter, J.A. 26, 48–9, 229–30
Schumpeterian: hypothesis 59,
    paradox 97, relations 54
Schwartz, N.L. 17, 39, 49
science 11–13
Scott, J.T. 49
selection environment 27
Simon, H.A. 17, 69, 74
Slama, J. 97, 194
social benefits 38–9
social optimum 25–6
socialist aims 98
socially managed industry 49, 55–7,
    62–4, *see also* centrally managed
    economies (CMEs)
Solow, R.M. 23, 76
Soviet Union, capital-to-output
    ratio 158; economic
    performance 95–6; efficiency
    96–7; foreign trade 105–6;
    R&D 94–5; rate of diffusion
    109; reforms 104, 106, 112
spill-over effect 51–5, 59, 62–4
Stafford, A. 12
Stewart, M. 69–70